T0257700

Hydraulic Fracturing

Volume III

Hydraulic Fracturing
Volume III

Edited by **Alexis Federer**

New York

Published by Callisto Reference,
106 Park Avenue, Suite 200,
New York, NY 10016, USA
www.callistoreference.com

Hydraulic Fracturing: Volume III
Edited by Alexis Federer

International Standard Book Number: 978-1-63239-424-8 (Hardback)

Contents

Preface

The objective of this book is to upgrade hydraulic fracturing technology that is effective in its purpose and sustainable in its impacts on communities and environments by bringing together hydraulic fracturing experts not only from the petroleum industry, but also from other application areas of hydraulic fracturing such as mining and geothermal energy production. Some of the discussed topics are Numerical Modeling, Mining and Measurement, Thermo-Hydro-Mechanical Systems, Experimental Geomechanics and Optimizing Stimulation of Fractured Reservoirs.

Various studies have approached the subject by analyzing it with a single perspective, but the present book provides diverse methodologies and techniques to address this field. This book contains theories and applications needed for understanding the subject from different perspectives. The aim is to keep the readers informed about the progresses in the field; therefore, the contributions were carefully examined to compile novel researches by specialists from across the globe.

Indeed, the job of the editor is the most crucial and challenging in compiling all chapters into a single book. In the end, I would extend my sincere thanks to the chapter authors for their profound work. I am also thankful for the support provided by my family and colleagues during the compilation of this book.

Editor

Numerical Modeling 2

Testing and Review of Various Displacement Discontinuity Elements for LEFM Crack Problems

Hyunil Jo and Robert Hurt

Additional information is available at the end of the chapter

Abstract

The numerical modeling of hydraulic fractures in unconventional reservoirs presents significant challenges for field applications. There remains a need for accurate models that field personnel can use, yet remains consistent to the underlying physics of the problem [1]. For numerical simulations, several authors have considered a number of issues: the coupling between fracture mechanics and fluid dynamics in the fracture [2], fracture interaction [3-5], proppant transport [6], and others [7-9]. However, the available literature within the oil and gas industry often ignores the importance of the crack tip in modeling applications developed for engineering design. The importance of accurate modeling of the stress induced near the crack tip is likely critical in complex geological reservoirs where multiple propagating crack tips are interacting with natural fractures. This study investigates the influence of various boundary element numerical techniques on the accuracy of the calculated stress intensity factor near the crack tip and on the fracture profile, in general. The work described here is a part of a long-term project in the development of more accurate and efficient numerical simulations for field engineering applications.

For this investigation, the authors used the displacement discontinuity method (DDM). The numerical technique is applied using constant and higher-order elements. Further, the authors also applied special crack tip elements, derived elsewhere [10], to capture the square root displacement variation at the crack tip, expected from Linear Elastic Fracture Mechanics (LEFM). The authors expect that special crack tip elements will provide the necessary flexibility to choose other tip profiles. The crack tip elements may prove instrumental for efficient modeling of the different near-tip displacement profiles exhibited by Viscosity-Dominated or Toughness-Dominated regimes in hydraulic fracture propagation. As others have shown [1,4,7], the accuracy of tip asymptote is critical in characterizing the stresses in the near-tip region of a propagating fracture.

The authors examined the numerically derived stress intensity factor for several crack geometries with and without higher-order elements and with and without special tip elements, to analytical solutions. As expected, they found that the cases with higher-order elements and special tip elements provide more accurate results than the cases with constant displacement discontinuity and/or no tip elements. However, the numerical technique developed still proved efficient.

These results show that numerical simulators can incorporate proper crack-tip treatments effectively. In addition, higher-order elements increase computational efficiency by reducing the number of elements instead of simply increasing the discretization of constant displacement elements. The accurate modeling of stress intensity factors is necessary to better simulate curved fractures, kinked cracks and interaction between fractures.

Keywords displacement discontinuity method, higher order elements, crack tip elements

1. Introduction

As new energy sources are sought for economic and security reasons, unconventional reservoirs attracted the oil and gas industry's attention. Among the unconventional options, shale gas reservoirs have become conspicuous. It is generally accepted that horizontal drilling and hydraulic fracturing are required to effectively recover hydrocarbons from the shale reservoirs [11]. Creating complex fracture networks by hydraulic fracturing is one of the most efficient ways to produce hydrocarbons from these reservoirs due to very low effective permeability (~500 nano Darcy). However, the numerical modeling of hydraulic fractures in such low permeable reservoirs presents significant challenges in field applications [1].

There remains a need for fast, yet accurate, models that remain consistent to the underlying physics of the problem. For numerical simulations, several researchers have considered a number of issues: the coupling between fracture mechanics and fluid dynamics in the fracture [2], fracture interaction [3-5], proppant transport [6], and others [7-9]. Further, there have been specific codes developed to model complex fracture network development [14-16]. Nevertheless, the available literature within the oil and gas industry often ignores the significance of the crack tip in modeling applications developed for hydraulic fracture design.

The importance of accurate modeling of the stress induced near the crack tip is likely critical in complex geological reservoirs. Multiple propagating crack tips interact with each other along with natural fractures, discontinuities, etc., during stimulation treatments in these reservoirs. Consequently, accurate modeling of the stress ahead of the propagating fracture is required to predict fracture paths in this complex environment. This study investigates the influence of several boundary element numerical techniques, available in the literature [10,12,13], on the stress intensity factor near the crack tip and on the fracture profile, in general.

This work is a part of a long-term project in the development of more accurate and efficient numerical simulations for field engineering applications.

To perform the investigation, we used the displacement discontinuity method (DDM), a version of the boundary element method (BEM). The method was developed for, and has been successfully applied to rock mechanics area such as mining engineering [17,18], fracture analysis [19,20], and wellbore stabilities [12]. We have applied DDM here using both constant and higher-order elements. The higher-order elements use a quadratic variation of displacement discontinuity, and are based on the use of three collocation points over a three-element patch centered at the source element [10], while the constant elements use a constant variation of displacement discontinuity [12]. Details related to the elements are elaborated on in Shou's work [12]. Further, the authors also applied special crack tip elements [10] to capture the square root displacement variation at the crack tip, expected from Linear Elastic Fracture Mechanics (LEFM). The authors expect that special crack tip elements will provide the necessary flexibility to choose other tip profiles. This flexibility will be instrumental for efficient modeling of the different near-tip displacement profiles exhibited by various regimes in hydraulic fracture propagation (e.g., Viscosity-Dominated or Toughness-Dominated [22,23]). As others have shown, the accuracy of tip asymptote is critical in characterizing the stresses in the near-tip region of a propagating fracture [1].

We examined the numerically derived stress intensity factor for three crack geometries with and without higher-order elements and with and without special tip elements, to analytical solutions. The three crack geometries are a pressurized crack orthogonal to the least principle stress, a slanted straight crack, and a circular arc crack. Several authors selected these specific geometries to justify the use of higher-order or specialized boundary elements [24-27]. However, the quantification of the computational efficiency coupled with the accuracy has been limited. Therefore, we present the following analysis that aids in determining the method that provides the most efficient, yet accurate solutions. Accurate and efficient methods are required for the development of field applications of engineering software packages.

Several other numerical techniques can be implemented within BEM. What we present here is not meant to be a review of possible combinations. We have chosen basic numerical techniques that provide the necessary flexibility to model very complex geometries, yet remain efficient enough for engineering modeling applications. The literature contains numerous examples of refinements to the techniques presented here [24-27]. For example, refinements with respect to the quarter-point method are found in Gray *et al.* [26] and refinements to higher-order elements are suggested by Dong and de Pater [25]. It is expected that implementing more refined methods will increase the efficiency of the numerical calculations. However, this work is primarily concerned with determining the general framework for BEM implementation.

The details related to the crack tip elements are available from a number of sources [10,12]. For brevity, this work will only summarize some basic concepts and mathematical formulas of the higher-order elements and the specialized crack tip elements. The next section of this paper describes the general displacement discontinuity method utilizing constant displacement elements. In section 3, the authors summarize the chosen higher order elements. Section 4 defines the special crack tip elements used in this work. Section 5 compares various combinations of the presented methods to known solutions of various crack geometries for an estimation of accuracy of calculations. Section 5 concludes by comparing of the computational efficiencies exhibited by the various methods. Finally, some concluding remarks are provided in Section 6.

2. Displacement discontinuity method

The displacement discontinuity method (DDM), originally formulated by Crouch [12], is used here. DDM is based on the solution of the stresses and displacements at a point caused by a constant displacement discontinuity (DD) over a line segment in an elastic body under prescribed boundary conditions [12]. Due to the simplicity of mathematical formulas and procedures of DDM (with a constant DD), it has been widely applied to various engineering problems. This paper summarizes some of the important mathematical expressions but limits specificity. The details of DDM are well described in the literature [12].

The 2-D displacements and stresses at a point (x, y), generated by a displacement discontinuity $(D_x(x), D_y(x))$ on the line segment $|x| \leq a$, $y=0$, can be analytically expressed as follows [1-3]:

$$u_x = [2(1-v)f_{,y} - yf_{,xx}] + [-(1-2v)g_{,x} - yg_{,xy}] \tag{1}$$

$$u_y = [(1-2v)f_{,x} - yf_{,xy}] + [2(1-v)g_{,y} - yg_{,yy}] \tag{2}$$

$$\sigma_{xx} = 2G[2f_{,xy} + yf_{,xyy}] + 2G[g_{,yy} + yg_{,yyy}] \tag{3}$$

$$\sigma_{yy} = 2G[-yf_{,xyy}] + 2G[g_{,yy} - yg_{,yyy}] \tag{4}$$

$$\sigma_{xy} = 2G[f_{,yy} + yf_{,yyy}] + 2G[-yg_{,xyy}] \tag{5}$$

where $f(x, y)$ and $g(x, y)$ are defined as:

$$f(x,y) = \frac{-1}{4\pi(1-v)} \int_{-a}^{a} D_x(\xi) \ln[\sqrt{(x-\xi)^2 + y^2}] d\xi \tag{6}$$

$$g(x,y) = \frac{-1}{4\pi(1-v)} \int_{-a}^{a} D_y(\xi) \ln[\sqrt{(x-\xi)^2 + y^2}] d\xi \tag{7}$$

and where the displacement discontinuity components are defined as:

$$D_x(x) = u_x(x,0^-) - u_x(x,0^+) \tag{8}$$

$$D_y(x) = u_y(x,0^-) - u_y(x,0^+) \tag{9}$$

For constant displacement elements (i.e. constant $D_x(x)$ and $D_y(x)$), the DD components can come out of the integrals, and then Equations (6) and (7) can be simplified as:

$$f(x,y) = I_0(x,y)D_x \tag{10}$$

$$g(x,y) = I_0(x,y)D_y \tag{11}$$

where

$$\begin{aligned}
I_0(x,y) &= \int_{-a}^{a} \ln[\sqrt{(x-\xi)^2 + y^2}]d\xi \\
&= y\left[\arctan\frac{y}{(x-a)} - \arctan\frac{y}{(x+a)}\right] - (x-a)\ln\sqrt{(x-a)^2 + y^2} \\
&\quad + (x+a)\ln\sqrt{(x+a)^2 + y^2} - 2a
\end{aligned} \tag{12}$$

For simplicity, the derivatives of $I_0(x, y)$, used to calculate the stresses and displacements (i.e. Equations (1) to (5)), are omitted in this paper. The derivatives are given in Shou *et al.* [10]. Since the numerical procedures of DDM (with constant displacement discontinuities) are well established in the available literature [12], they are not given herein.

However, DDM with a constant DD can't accurately calculate the stresses and displacements of the area closer than about one element-length distance from a boundary [12]. To improve the accuracy of calculations in close proximity to the boundaries, Crawford *et al.* developed higher-order displacement elements [24] among others [25]. Although higher-order elements overcame the limitations of constant elements and improved the accuracy of DDM, the method significantly increases the number of degrees of freedom. In other words, the higher-order elements increase the number of equations that must be solved.

To improve the accuracy of DDM without sacrificing the number of degrees of freedom of the overall system, a new higher-order elements method was suggested by Shou *et al.* [10]. The method used collocation points at the centers of the source elements and its two adjacent neighbors, so it could maintain the same degrees of freedom as the constant elements method by sharing the DD of the two adjacent neighbors. Other methods have been suggested by in the literature [15,25] to overcome issues with kinked or intersecting cracks when utilizing neighboring elements in calculations.

This study uses Shou *et al.*'s method to satisfy one of this research's objectives, which is to develop methods that reduce computation costs while improving accuracy. The next section will summarize some basic concepts and mathematical formulas for the higher-order elements used in this work. As above, further details of the higher-order elements are available in [10].

3. Higher-order elements of displacement discontinuity method

Higher-order elements, as formulated by Shou *et al.*, use quadratic displacement elements. The calculation of the DD component of a particular element is accomplished by using three collocation points. The center collocation point is within the element of interest, while the bounding collocation points are within the neighboring elements. This configuration forms a three-element "patch" (shown in Figure 1), on which the quadratic formulation is performed. Equation (13) shows how the value of the DD components is formed mathematically.

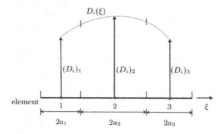

Figure 1. Quadratic collocation for the new higher-order elements [2]

$$D_i(\xi) = N_1(\xi)(D_i)_1 + N_2(\xi)(D_i)_2 + N_3(\xi)(D_i)_3 \tag{13}$$

where $(D_i)_1$, $(D_i)_2$ and $(D_i)_3$ are the nodal displacement discontinuities ($i = x$ or y) and

$$N_1 = \frac{\xi(\xi - a_2 - a_3)}{(a_1 + a_2)(a_1 + 2a_2 + a_3)} \tag{14}$$

$$N_2 = \frac{-(\xi + a_1 + a_2)(\xi - a_2 - a_3)}{(a_1 + a_2)(a_2 + a_3)} \tag{15}$$

$$N_3 = \frac{\xi(\xi + a_1 + a_2)}{(a_2 + a_3)(a_1 + 2a_2 + a_3)} \tag{16}$$

N_1, N_2 and N_3 are the collocation shape functions whose a_1, a_2 and a_3 are half length of the three elements of the patch.

Combining Equations (13) through (16) with Equations (6) and (7) gives the following simplified expressions:

$$f(x,y) = \frac{-1}{4\pi(1-v)}\sum_{j=1}^{3}(D_x)_j F_j(I_0, I_1, I_2)$$

(17)

$$g(x,y) = \frac{-1}{4\pi(1-v)}\sum_{j=1}^{3}(D_y)_j F_j(I_0, I_1, I_2)$$

(18)

where the subscript j indicates the j th collocation node in the three-element patch and $F_j(I_0, I_1, I_2)$ is defined as

$$F_j(I_0, I_1, I_2) = \int N_j(\xi)\ln[\sqrt{(x-\xi)^2 + y^2}]d\xi, \quad j = 1\,to\,3$$

(19)

which can be expressed in terms of constant, linear, and quadratic kernels (I_0, I_1, I_2). The definition of these kernels is given by Shou *et al.* [10].

Based on these formulas, a crack can be discretized into N elements (see Figure 2) and 2N equations in terms of the DD component unknowns are formed (i.e. 2N unknowns of D_x and D_y). Under certain boundary conditions, the 2N unknowns can be obtained. Once the 2N unknowns are calculated, the 2-D displacements and stresses at a point (x, y) can be calculated through Equations (1) to (5). Further, Equations (20) and (21) compute the stress intensity factors at the crack tip.

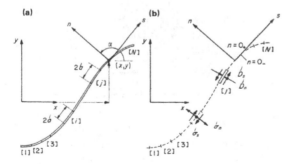

Figure 2. Representation of a crack by N elemental displacement discontinuities [12]

$$K_I = \frac{G}{4(1-v)}\sqrt{\frac{2\pi}{a^*}}D_n^*$$

(20)

$$K_{II} = \frac{G}{4(1-\nu)}\sqrt{\frac{2\pi}{a^*}}D_s^* \tag{21}$$

where a^* is the half length of the crack tip element, and D_n^* and D_s^* are normal and shear DD at the crack tip, respectively.

4. Crack tip elements of displacement discontinuity method

In addition to advanced elements, Shou *et al.* formulated two special crack tip elements to capture the square root displacement variation at the crack tip, expected from Linear Elastic Fracture Mechanics (LEFM) [10]. One is to use a constrained collocation point one-quarter of an element length away from the end of the crack. This study will refer to this as a quarter element method. The other tip element is simply a prescribed displacement discontinuity proportional to the square root variation at the crack tip. Herein, it is called the square root element method. This study uses their methods to calculate the stresses and displacements at the crack tip.

4.1. Quarter element method

Shou *et al.* introduced a constrained collocation point one-quarter of the crack tip element length away from the crack tip element. The DDs of the point will be set zero. Figure 3 shows the element.

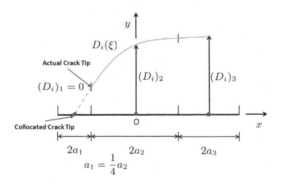

Figure 3. Quarter element method at a crack tip [2]

Numerical implementation of this method is more efficient compared to the square root element method and provides reasonably accurate results. However, the displacement discontinuities near the crack tip do not capture the square root displacement variation at the crack tip, expected from LEFM [10]. Theoretically, this method may give unreliable results.

Thus, Shou *et al.* introduced a more sophisticated crack tip elements method to comply with the LEFM physical phenomenon, which is the square root element method summarized below.

4.2. Square root element method

LEFM predicts that in the vicinity of the crack tip the crack displacement is proportional to the square root of the distance from the crack tip (i.e. $w \propto \sqrt{\xi}$). Figure 4 illustrates the basics of the square root crack tip element. Equation (22) shows the representation of $\sqrt{\xi}$ variation of the displacement discontinuities $D_i(\xi)$ along the crack tip.

$$D_i(\xi) = D_{ci}\sqrt{\frac{\xi}{a}} \quad i = x, y \tag{22}$$

where D_{ci} are the DD values at the center of the crack tip element. Substituting Equation (22) into Equations (1) to (5) the stresses and displacements are resolved in terms of D_{ci}. The solutions can be expressed in kernel functions, similar to higher-order elements methods. The details of the kernel functions were well documented in previous work [10].

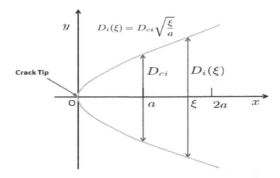

Figure 4. Square root crack tip element [2]

5. Comparison

To access the accuracy of the selected numerical techniques, this study compares the stress near the crack tips and stress intensity factors for three crack geometries with constant displacement or higher-order elements, both elements will be combined with and without specialized quarter and square root crack tip elements. We chose these particular geometries because analytical solutions are readily available. Further, many previous publications comparing BEMs have chosen these same geometries [25-27]. This research uses following elastic properties: $E = 10^6 \, psi$ and $v = 0.2$ in the calculations.

5.1. Single pressurized crack in an infinite elastic domain

A single pressurized crack is a basic fracture geometry, and the analytical solutions are well documented [19,28]. Figure 5 shows a schematic diagram for the fracture geometry. The crack is pressurized by a pressure $p=1000$ *psi*. The crack length is 10 inches and is discretized into 10 elements with equal length. According to the specified method, the two crack tips may be replaced by specialized crack tip elements. This research compares half width, stresses at defined locations, and stress intensity factors computed from each method.

Figure 5 illustrates the chosen locations where each of the given methods calculates the stress. The points are arbitrary, but chosen at a location of symmetry with respect to the fracture. In Figure 5, the blue X represents the point orthogonal to the fracture plane at the mid-point of the fracture. The red diamond represents a point ahead of the fracture tip. For convenience, this report uses the following abbreviations to represent each method: AM (analytical method), CDD (only constant displacement discontinuity), HDD (only higher-order elements), CDDCE (constant DD with the quarter element method), CDDCT (constant DD with the square root element method), HDDCE (higher-order elements with the quarter element method), and HDDCT (higher-order elements with the square root element method).

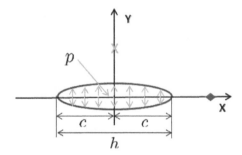

Figure 5. A crack under a constant pressure. The blue X represents a point orthogonal to the fracture plane where the induced stress calculated from each method is compared. The red diamond represents the evaluation point ahead of the fracture tip.

Equation (23) is the analytical solution of the dimensionless half width under a constant pressure [29].

$$\frac{w(x)}{c} = \frac{4p}{E'}\sqrt{1-\left(\tfrac{x}{c}\right)^2} \tag{23}$$

Figure 6 is a plot of the calculated fracture profile from each method. The analytical solution is the solid blue line. The computed half widths from each of the numerical methods are shown as a series of points. From the results, CDD overestimates the half width particularly near crack tip area as others have found [20]. Conversely, HDD underestimates the fracture width in the proximity of the tip. Methods that use tip elements show fracture width profiles close to the analytical solution. The importance of utilizing special crack tip elements is well established in the literature [10, 26, 27].

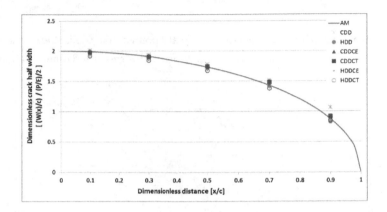

Figure 6. Dimensionless crack half width versus Dimensionless distance from the crack center (HDD, CDDCE, CDDCT and HDDCE overlap)

Figure 7 is more illustrative for comparing the accuracy of the various methods. It shows the relative error from the computation of the half width compared to the analytical solution (i.e. $\frac{w - w_{AM}}{w_{AM}}\%$). The relative errors of all methods increase as they approach to the crack tip. The majority of the methods demonstrate errors bounded between -5% to 5%, except for CDD, which shows over 20% in close proximity to the crack tip. Computational errors over 20% from CDD methods have been reported in the literature [20].

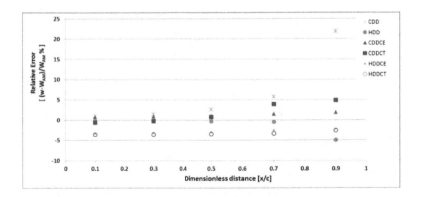

Figure 7. Relative error of the half width

To evaluate the perturbed stress state due to the presence of a pressurized crack we use [30]

$$\frac{\sigma_{xx}}{-p} = \frac{\left(\frac{x}{c}\right)}{\sqrt{\left(\frac{x}{c}\right)^2 - 1}} - 1 \tag{24}$$

Equation (24) provides a dimensionless stress $\frac{\sigma_{xx}}{p}$ at a point along X-axis (the distance from the crack tip normalized by the crack half-length), which in this case is red diamond in Figure 5.

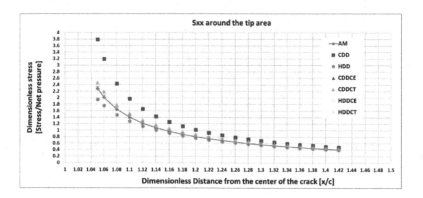

Figure 8. Dimensionless stress versus Dimensionless distance ahead of the crack tip (i.e. at red diamond) (CDDCE, HDDCE and HDDCT overlap)

Figure 8 plots the dimensionless stress ahead of the crack tip. x/c = 1 represents the crack tip. Figure 9 plots the relative stress (the ratio of $\frac{\sigma_{xx}}{p}$ to the analytical $\frac{\sigma_{xx}}{p}$) near the crack tip area for each numerical method.

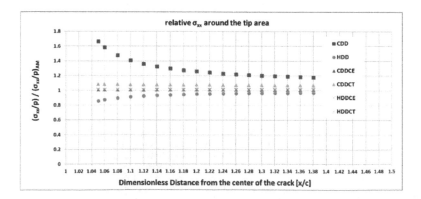

Figure 9. Relative stress versus Dimensionless distance ahead of the crack tip (CDDCE, HDDCE and HDDCT overlap)

These figures also show that CDD overestimates the stress and HDD underestimates it while the other methods give results with less than 1% error. The inaccuracies of the methods without tip elements become significant closer to the crack tip. Similar to the half width results, the results of the methods with tip elements overlap since they are close to the analytical solution.

To calculate the stress induced orthogonal to a pressurized crack we use [30]

$$\frac{\sigma_{yy}}{p} = -\frac{\left(\frac{L}{c}\right)}{\sqrt{\left(\frac{L}{c}\right)^2+1}} + 1 + \frac{\left(\frac{L}{c}\right)}{\left\{\sqrt{\left(\frac{L}{c}\right)^2+1}\right\}^3} \tag{25}$$

Equation (25) expresses a dimensionless stress $\frac{\sigma_{yy}}{p}$ at a point along the Y-axis (the distance from the crack face is normalized by the crack half-length), which is represented by the blue X in Figure 5. Figure 10 shows the results of the calculation from each numerical method. These results show consistency with the analytical solution, regardless of the numerical method. This is expected, as the location where the stress is calculated is sufficiently far from the crack tip, i.e. more than the length of one discretized element [12].

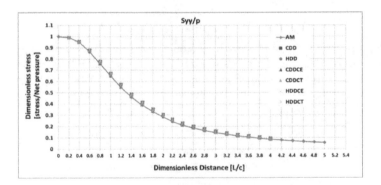

Figure 10. Dimensionless stress versus Dimensionless distance orthogonal to the fracture plane, i.e. at the blue X. In this case, all methods overlap

Table 1 shows the calculated stress intensity factor, along with the ratio to the analytical solution. For the mode I (or K_I), CDD shows the biggest error while CDDCE, HDDCE and HDDCT give around 1% errors. Obviously, the mode II (or K_{II}) stress intensity factor is zero. So, Table 1 omits the results.

Reasonable values from calculations of the stresses and displacements can be achieved at distances greater than the length of one discretized element, regardless of the numerical technique. Close to the crack tip, however, the error of displacements, stresses, and stress intensity factors for CDD elements are significant, whereas CDDCE, HDDCE and HDDCT provide reasonable estimations. This is not surprising; similar results are well documented in the literature [12,24,26].

	KI [psi √in]	KI/ KI$_{AM}$
AM	3963.3	1
CDD	4905.6	1.238
HDD	3670.8	0.926
CDDCE	3933.8	0.993
CDDCT	4219	1.065
HDDCE	3933.8	0.993
HDDCT	3918.5	0.989

Table 1. Stress intensity factors

5.2. A slanted straight crack

While a straight pressurized crack shows zero K_{II}, a slanted straight crack under a uniform tension can show a variable K_{II} depending on the angle of incidence to the applied tension. Figure 11 illustrates the crack geometry. The stress intensity factors are calculated by [19]

$$K_I = \sigma\sqrt{\pi a}\sin^2(\beta) \qquad K_{II} = \sigma\sqrt{\pi a}\sin(\beta)\cos(\beta) \tag{26}$$

Equation (26) expresses the analytical solution of the stress intensity factors [19]. The uniform tension is $\sigma = 1000$ psi. The crack length is 10 inches. It is discretized into 10 elements with equal length. The two crack tips are replaced by crack tip elements according to the applied method.

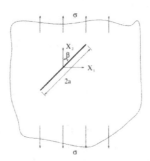

Figure 11. Slanted straight crack under uniform axial tension at infinity [25]

Figure 12 shows the dimensionless K_I (i.e. $\dfrac{K_I}{\sigma\sqrt{\pi a}}$) according to the slanted angle and Figure 13 gives the dimensionless K_{II} (i.e. $\dfrac{K_{II}}{\sigma\sqrt{\pi a}}$) according to the slanted angle, respectively. Similar to the previous crack geometry, CDD and CDDCT overestimate K_I and K_{II} while HDD underestimates them. CDDCE, HDDCE, and HDDCT show fairly accurate results, so that they overlap in Figure 13 and 14. However, the calculation errors of the two stress intensity factors exhibit opposing patterns. For K_I, the errors increase as the slanted angle becomes larger. Conversely, the errors of K_{II} are maximal at 45° and at a minimum at 0° and 90°.

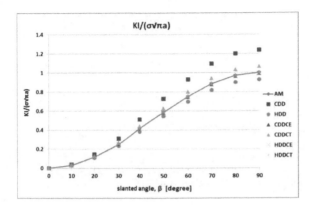

Figure 12. Dimensionless K_I versus the slanted angle (CDDCE, HDDCE and HDDCT overlap)

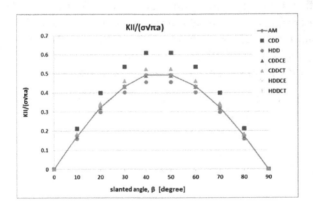

Figure 13. Dimensionless K_{II} versus the slanted angle (CDDCE, HDDCE and HDDCT overlap)

5.3. A circular arc crack

Curved cracks may represent more realistic fracture geometry and exhibit complexity in calculation of the stress intensity factors. This study selects a circular arc crack under a far field uniform biaxial tension in order to evaluate the accuracy of the numerical methods. Figure 14 describes the fracture geometry. The uniform biaxial tension is $\sigma = 1000 \ psi$. The analytical solutions of K_I and K_{II} for a curved crack are given by [19]

$$K_I = \frac{\sigma\sqrt{\pi r \sin(\alpha/2)}}{1+\sin^2(\alpha/4)}\cos(\alpha/4) \qquad K_{II} = \frac{\sigma\sqrt{\pi r \sin(\alpha/2)}}{1+\sin^2(\alpha/4)}\sin(\alpha/4) \tag{27}$$

where r is the radius of the circular arc [19].

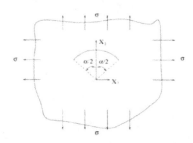

Figure 14. Circular arc crack under uniform biaxial tension [8]

Figures (15) and (16) show the values of the calculated K_I and K_{II} as a function of the circular crack angle, respectively. The figures depict actual stress intensity factor values under the prescribed biaxial tension, instead of dimensionless values. The unit of K_I and K_{II} is $psi\sqrt{in}$. The effective half-length of the crack is ambiguous due to variable circular arc length related to the prescribed circular arc angle (α). The circular arc has a 10 inch radius. It is discretized into 20 elements with equal length. The two tip elements are evenly discretized into 10 additional segments to apply the tip elements methods, respectively. Thus, the circular arc has 38 segments (i.e. 18 middle identical elements and 20 identical tip elements).

Figures (17) and (18) show the normalized K_I and K_{II} stress intensity factors as a function of the circular crack angle, respectively.

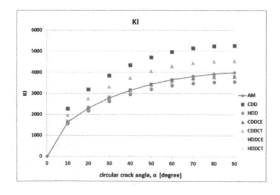

Figure 15. KI versus the circular arc angle (CDDCE, HDDCE and HDDCT overlap)

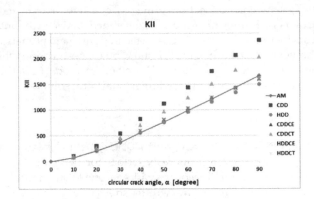

Figure 16. KII versus the circular arc angle (CDDCE, HDDCE and HDDCT overlap)

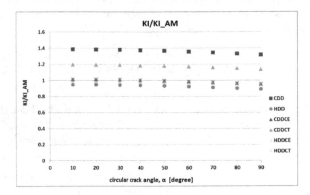

Figure 17. Relative K_I versus the circular arc angle (CDDCE, HDDCE and HDDCT overlap)

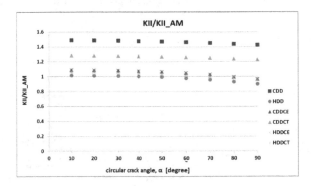

Figure 18. Relative K_{II} versus the circular arc angle (CDDCE, HDDCE and HDDCT overlap)

Based on the results presented in this section, numerical calculations of K_I and K_{II} show similar patterns. As the circular crack angle increases, the differences between the analytically derived stress intensity factors increase. The ratio between the analytical and numerical stress intensity factors decreases, however. CDD and CDDCT show the overestimation while HDD method underestimates the stress intensity factors. The other methods (CDDCE, HDDCE, and HDDCT) provide very close results compared to the analytical solution.

5.4. Computational efficiency

In general, we find that CDDCE, HDDCE, and HDDCT methods significantly increase the accuracy of the computation of stress intensity factors for the geometries presented here. However, the required computational resource varies among the numerical methods. Further, for constant displacement elements, accuracy of the stress calculations ahead of the crack tip can be improved by increasing the number of elements. In order to objectively evaluate the efficiency of the numerical methods, we return to the pressurized crack example from Section 5.1. The following section first compares the accuracy improvement by increasing the number of elements for the constant element method. Then we compare the computation time for calculating the stress intensity factors for this particular problem.

The computer specifications used in this work are as follows: CPU-Intel® Xeon W3670 @ 3.2GHZ, installed memory (RAM)-24 gigabytes, OS- 64-bit Windows 7®, Software- Matlab® R2011b.

This study uses the stress of a horizontal crack near the crack tip to show the computational accuracy of simply increasing the number of CDD elements compared to higher-order elements and/or special tip elements. Figure 19, shows the normalized stress (the ratio of σ_{xx} / p to the analytical σ_{xx} / p) near the crack tip as a function of the number of CDD elements. The number at the legend indicates the number of CDD elements. As expected, increasing the number of elements results in a corresponding improvement in calculation accuracy. Figure 19 illustrates that at least 100 CDD elements are required to show less than 1% error. In other words, an order of magnitude increase in the number of CDD elements is required to match the accuracy derived with CDDCE, HDDCE and HDDCT methods (see Figure 9).

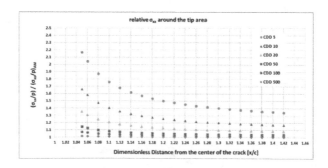

Figure 19. Relative stress versus Dimensionless distance ahead of the crack tip (i.e. at red diamond) according to the number of CDD elements used

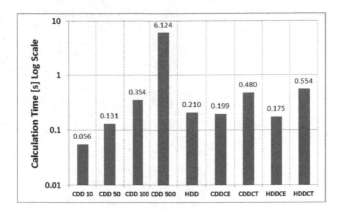

Figure 20. Computation time for the solution of the pressurized crack exercise

Finally, the computational efficiency is evaluated by inspecting Figure 20. This figure plots the calculation time of the various numerical methods for the pressurized crack exercise described above. As expected, CDD calculations with fewer elements are completed more quickly. Interestingly the CDDCE, HDDCE, and HDDCT methods show similar computation times while maintaining the highest accuracy of the numerical methods. This result allows for further refinement and evaluation of the CDDCE, HDDCE, and HDDCT methods using more refined element choices and increasingly complex crack geometries.

6. Conclusion

Overall results show that CDD gives prominent errors of calculations of stresses, displacements, and stress intensity factor compared to the other methods. Particularly, when approaching to the crack tips and a fracture is curved, the errors of CDD significantly increase. Replacing tip elements by special crack tip elements can mitigate calculation errors when close to the crack tips. Using higher-order elements helps to reduce errors for the simple straight crack geometries. When a fracture is curved, the efficiency of combining specialized crack tip elements in computational errors in the calculation of K_I and K_{II} is more important than for simple fracture geometries. Combination of higher-order elements and crack tip elements give the most accurate calculations, yet retain the necessary efficiency. However, the overall efficiency of CDDCE, HDDCE and HDDCT methods cannot be definitively evaluated using the simple geometries shown here. We reserve that analysis for subsequent publications.

For comparison within this work, the numerical methods maintained a similar number of elements for each fracture geometry. Increasing the number of CDD elements increases the accuracy of the CDD method. However, this requires increased computation time. Thus, the use of higher-order elements and crack tip elements is likely warranted if considering the development of more accurate and efficient numerical simulations for field engineering applications where computation resources are restricted. Evaluating the efficiency of specific combinations of higher-order elements coupled with specialized crack tip elements requires more complex geometries than presented here.

Nomenclature

u	Displacement
σ	Stress
E	Young's modulus
v	Poisson's ratio
G	Shear modulus
E'	Plane strain, $\frac{E}{1-v^2}$
c, a	Fracture half-length or half height
h	Fracture length or height

Acknowledgements

The authors thank Baker Hughes for supporting this research and for permission to publish this paper, Randy LaFollette, for sincere advice and encouragement, and Russell Maharidge, PhD for considerate reviews.

Author details

Hyunil Jo* and Robert Hurt

*Address all correspondence to: hyunil.jo@bakerhughes.com

Baker Hughes, Tomball, TX, USA

References

[1] Adachi, A, Siebrits, E, Peirce, A, & Desroches, J. Computer simulation of hydraulic fractures. International Journal Of Rock Mechanics And Mining Sciences (2007). , 44(5), 739-757.

[2] Naceur, K. B, Thiercelin, M, & Touboul, E. Simulation of Fluid Flow in Hydraulic Fracturing: Implications for 3D Propagation. SPE Production Engineering (1990). SPE 16032, 5(2), 133-141.

[3] Germanovich, L, & Astakhov, D. Multiple fracture model. In: Jeffrey, McLennan, editors. Proceedings of the three dimensional and advance hydraulic fracture modeling. Workshop held at 4th North American rock mechanics symposium, Seattle, July 29-31, (2000). , 45-70.

[4] Bunger, A. P, Zhang, X, & Jeffrey, R. G. Parameters Affecting the Interaction Among Closely Spaced Hydraulic Fractures. SPE Journal;17(1):SPE-140426-PA, 292-306.

[5] Olson, J. E, & Wu, K. Sequential vs. Simultaneous Multizone Fracturing in Horizontal Wells: Insights From a Non-Planar, Multifrac Numerical Model. In: SPE Hydraulic Fracturing Technology Conference. The Woodlands, Texas, USA: Society of Petroleum Engineers. SPE-M, 152602.

[6] Gu, H, & Siebrits, E. On numerical solutions of hyperbolic proppant transport problems. In: Proceedings of the 10th international conference on hyperbolic problems: theory, numerics and applications, Osaka, Japan, September (2004). , 13-17.

[7] Gordeliy, E, & Detournay, E. A fixed grid algorithm for simulating the propagation of a shallow hydraulic fracture with a fluid lag. International Journal for Numerical and Analytical Methods in Geomechanics;35(5):602.

[8] Zhang, X, Jeffrey, R. G, & Thiercelin, M. Deflection and propagation of fluid-driven fractures at frictional bedding interfaces: A numerical investigation. Journal of Structural Geology (2007).

[9] Sellers, E, & Napier, J. A comparative investigation of micro-flaw models fog the simulation of brittle fracture in rock. Computational Mechanics (1997).

[10] Shou, K. J, & Crouch, S. L. A higher order displacement discontinuity method for analysis of crack problems. International Journal of Rock Mechanics and Mining Science & Geomechanics Abstracts (1995).

[11] King, G. E. Thirty Years of Gas Shale Fracturing: What Have We Learned? In: SPE Annual Technical Conference and Exhibition. Florence, Italy: Society of Petroleum Engineers. SPE 133456

[12] Crouch, S. L, & Starfield, A. M. Boundary Element Methods in Solid Mechanics: With Applications in Rock Mechanics and Geological Engineering: George Allen & Unwin; 1 edition, (1983).

[13] Crouch, S. L. Solution of plane elasticity problems by the displacement discontinuity method. I. Infinite body solution. International Journal for Numerical Methods in Engineering (1976).

[14] Olson, J. E. Predicting fracture swarms-- the influence of subcritical crack growth and the crack-tip process zone on joint spacing in rock. Geological Society, London, Special Publications (2004). , 231(1), 73-88.

[15] Wu, R, Kresse, O, Weng, X, Cohen, C-e, & Gu, H. Modeling of Interaction of Hydraulic Fractures in Complex Fracture Networks. In: SPE Hydraulic Fracturing Technology Conference. The Woodlands, Texas, USA: Society of Petroleum Engineers. SPE152052

[16] Weng, X, Kresse, O, Cohen, C-E, Wu, R, & Gu, H. Modeling of Hydraulic-Fracture-Network Propagation in a Naturally Fractured Formation. SPE Production & Operations;26(4):SPE-140253, 368-380.

[17] NapierJAL. Modelling of fracturing near deep level gold mine excavations using a displacement discontinuity approach. Mechanics of jointed and faulted rock. Rossmanith (ed), Balkema: Rotterdam: (1990). , 709-716.

[18] Peirce, A. P. Napier JAL. A Spectral Multipole Method For Efficient Solution Of Large-Scale Boundary-Element Models In Elastostatics. International Journal For Numerical Methods In Engineering (1995). , 38(23), 4009-4034.

[19] Tada, H, Paris, P. C, & Irwin, G. R. The Stress Analysis of Cracks Handbook. 3 ed: American Society of Mechanical Engineering; (2000).

[20] Exadaktylos, G, & Xiroudakis, G. A G2 constant displacement discontinuity element for analysis of crack problems. Computational Mechanics;, 45(4), 245-261.

[21] Shou, K-J. Napier JAL. A two-dimensional linear variation displacement discontinuity method for three-layered elastic media. International Journal of Rock Mechanics and Mining Sciences (1999).

[22] Garagash, D. I, & Detournay, E. Near tip processes of a fluid-driven fracture. ASME J Appl Mech (2000). , 67, 183-92.

[23] Detournay, E. (2004). Propagation regimes of fluid driven fractures in impermeable rocks. Int. J. Geomechanics , 4(1), 1-11.

[24] Crawford, A. M, & Curran, J. H. Higher-order functional variation displacement discontinuity elements. International Journal of Rock Mechanics and Mining Sciences & Geomechanics Abstracts (1982).

[25] Dong, C. Y, & De Pater, C. J. (2001). Numerical implementation of displacement discontinuity method and its application in hydraulic fracturing. Computer Methods in Applied Mechanics and Engineering;191:745.

[26] Gray, L. J, Phan, A. V, Paulino, G. H, & Kaplan, T. Improved quarter-point crack tip element. Engineering Fracture Mechanics (2003). , 70(2), 269-283.

[27] Yan, X. Q. A special crack tip displacement discontinuity element. Mechanics Research Communications (2004). , 31(6), 651-659.

[28] Pollard, D, & Segall, P. Theoretical displacements and stresses near fractures in rock: with applications to faults, joints, veins, dikes and solution surfaces. In: ATKINSON, B K. Fracture Mechanics of Rock. Academic Press, London, (1987). , 277-350.

[29] Valkó, P, & Economides, M. J. Hydraulic Fracture Mechanics. 1 ed: Wiley; (1995).

[30] Sneddon, I. N. The Distribution Of Stress In The Neighbourhood Of A Crack In An Elastic Solid. Proceedings Of The Royal Society Of London Series A-Mathematical And Physical Sciences (1946). , 187(1009), 229-260.

Chapter 2

Three-Dimensional Numerical Model of Hydraulic Fracturing in Fractured Rock Masses

B. Damjanac, C. Detournay, P.A. Cundall and Varun

Additional information is available at the end of the chapter

Abstract

Conventional methods for simulation of hydraulic fracturing are based on assumptions of continuous, isotropic and homogeneous media. These assumptions are not valid for most rock mass formations, particularly shale gas reservoirs, as these typically consist of a large volume of naturally fractured rock in which propagation of a hydraulic fracture (HF) involves both fracturing of intact rock and opening or slip of pre-existing discontinuities (joints). The pre-existing joints can significantly affect the HF trajectory, the pressure required to propagate the fracture and also the leak-off from the fracture into the surrounding formation. None of these effects can be simulated using conventional methods.

HF Simulator is a new three-dimensional numerical code that can simulate propagation of hydraulic fracture in naturally fractured reservoirs, accounting for the interaction between the hydraulic fracture and pre-existing joints. In *HF Simulator*, fracture propagation occurs as a combination of intact-rock failure in tension, and slip and opening of joints. The code uses a lattice representation of brittle rock consisting of point masses (nodes) connected by springs. The pre-existing joints are derived from a user-specified discrete fracture network (DFN).

HF Simulator can model fluid injection or production from one or multiple boreholes each with one or multiple clusters. Non-steady, hydro-mechanically coupled fluid flow and pressure within the network of joint segments and the rock matrix are considered.

An outline of the code hydro-mechanical formulation is presented and examples are provided to illustrate the code capabilities.

Keywords: Numerical model, naturally fractured, rock mass

1. Introduction

A new generation tool that uses the bonded particle model (BPM) [1] and the synthetic rock mass (SRM) concept [2] has been developed to model hydraulic fracture (HF) propagation in naturally fractured reservoirs (NFRs).

Most rock mass formations, and shale gas reservoirs in particular, consist of a large volume of fractured rock in which propagation of an HF involves both fracturing of intact rock and opening or slip of pre-existing discontinuities (joints). The pre-existing joints can significantly affect the HF trajectory, the pressure required to propagate the fracture, but also the leak-off from the fracture into the surrounding formation. None of these effects can be simulated using conventional hydraulic fracturing simulation methods, based on assumptions of continuous, isotropic and homogeneous media.

To address this challenge, a numerical approach called SRM method [2] has been developed recently based on the distinct element method. SRM method usually is realized as a bonded-particle assembly representing brittle rock containing multiple joints, each one consisting of a planar array of bonds that obey a special model, namely the smooth joint model (SJM). The SJM allows slip and separation at particle contacts, while respecting the given joint orientation rather than local contact orientations. Overall fracture of a synthetic rock mass depends on both fracture of intact material (bond breaks), as well as yield of joint segments.

Previous SRM models have used the general-purpose codes *PFC2D* and *PFC3D* [3,4], which employ assemblies of circular/spherical particles bonded together. Much greater efficiency can be realized if a "lattice," consisting of point masses (nodes) connected by springs, replaces the balls and contacts (respectively) of *PFC3D*. The lattice model still allows fracture through the breakage of springs along with joint slip, using a modified version of the SJM. The new 3D program, *HF Simulator* described in this paper, is based on such a lattice representation of brittle rock. *HF Simulator* overcomes all main limitations of the conventional methods for simulation of hydraulic fracturing in jointed rock masses and is computationally more efficient than *PFC*-based implementations of the SRM method.

The formulation of the code is described in this paper. The examples of code verification and application are also presented.

2. Model description

2.1. Background: Synthetic rock mass approach

Over past years, the SRM has been developed [2] as a more realistic representation of mechanical behavior of the fractured rock mass compared to conventional numerical models. The SRM consists of two components: (1) the bonded particle model (BPM) of deformation and fracturing of intact rock, and (2) the smooth joint model (SJM) of mechanical behavior of discontinuities.

The BPM, originally implemented in *PFC*, is created when the contacts between the particles (disks in 2D and spheres in 3D) are assigned certain bond strength (both in tension and shear). It was found that BPM quite well approximates mechanical behavior of the brittle rocks [1]. The elastic properties of the contacts (i.e., contact shear and normal stiffness) can be calibrated to match the desired elastic properties (e.g., Young's modulus and Poisson's ratio) of the assembly of the particles. Similarly, the tensile and shear contact strengths can be adjusted to match the macroscopic strengths under different loading conditions (e.g., direct tension, unconfined and confined compression).

In the BPM, the contact behavior is perfectly brittle. Breakage of the bond, a function of the forces in the contact and the bond strength, corresponds to formation of a microcrack. An example of unconfined compression test conducted using *PFC2D* is illustrated in Figure 1, which shows recorded axial stress-strain response and the model configuration with generated microcracks. The shear microcracks are black; the tensile microcracks are red. Shown is the state when the sample is loaded beyond its peak strength. The stress-strain curve exhibits characteristics typical of brittle rock response. For the load levels less than ~80% of the peak strength, the stress-strain response is linearly elastic, with the slope of the line equal to the Young's modulus. Some microcracks, randomly distributed within the sample, start developing at the load levels greater than ~40% of the peak strength. Significant non-linearity develops as the load exceeds 80% of the peak strength. In this phase, the microcracks begin to coalesce, forming fractures on the scale of the sample. After the peak strength is reached, the material starts to soften (i.e., to lose the strength). At this stage, as shown in Figure 1, the failure mechanism and the "shear bands" are well developed. It is interesting that in the unconfined compression test, the majority of cracks are tensile (red lines in Figure 1). The "shear bands" on the scale of the sample are formed by coalescence of a large number of tensile microcracks.

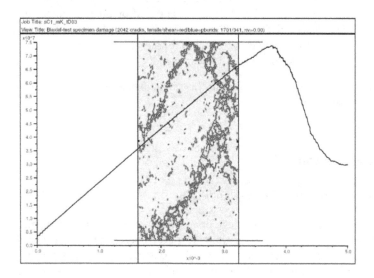

Figure 1. Example of unconfined compressive test using bonded particle model (BPM).

In order to model a typical rock mass in the BPM, it is also necessary to represent pre-existing joints (discontinuities). A straightforward approach is to simply break or weaken the bonds (in the contacts between the particles) intersected by the pre-existing joints. The created discontinuity will have roughness with the amplitude and wavelength related to the resolution, or the particle size of the BPM. The mechanical behavior of discontinuities is very much affected by their roughness. The problem is that the selected particle size (or resolution) typically is not related to actual roughness of the pre-existing joints. The SJM overcomes this limitation. The contacts in the BPM model are oriented in the direction of the line connecting the centers of the particles involved in the contact. The SJM contacts are oriented perpendicular to the fracture plane irrespective of the relative position of the particles. Consequently, the particles can slide relative to each other in the plane of the fracture as if it is perfectly smooth.

The SRM and its components are shown in Figure 2. The BPM represents the intact rock, its deformation and damage. The pre-existing joints are represented explicitly, using the SJM. They can be treated deterministically, by specifying each discontinuity by its position and orientation as mapped in the field. However, typically, for practical reasons, it is not possible to treat the DFN deterministically. Instead, fracturing in the rock mass is characterized statistically. The synthetic DFNs that are statistically equivalent (i.e., fracture spacing, orientation and size) to fracturing of the rock mass are generated and imported into the SRM using SJM (Figure 2). Very often a reasonable compromise is to represent few dominant structures (faults) with their deterministic position and orientation and the rest of the fracturing in the rock mass (smaller structures) using a synthetic DFN.

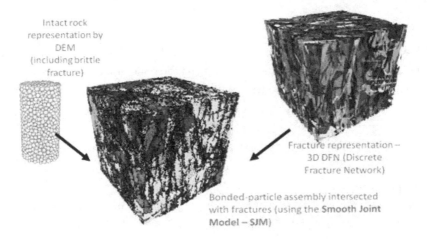

Intact rock representation by DEM (including brittle fracture)

Fracture representation – 3D DFN (Discrete Fracture Network)

Bonded-particle assembly intersected with fractures (using the Smooth Joint Model – SJM)

Figure 2. Synthetic rock mass (SRM).

One of the advantages of the SRM is that the components, the intact rock and the joints, can be mechanically characterized by standard laboratory tests. The mechanical response of the rock mass and the size effect are the model results, functions of the model size, DFN characteristics and mechanical properties of the components. Thus, it is not necessary to rely on empirical relations to estimate the rock mass properties and to account for the size effect considering the size of the samples tested in the laboratory and the scale of interest in the model.

The new code, *HF Simulator*, is based on implementation of the SRM in the lattice, which is a simplified, but also a computationally more efficient version of particle flow code (*PFC*). Despite simplifications, the lattice approach represents all physics important for simulation of hydraulic fracturing.

2.2. Lattice

The lattice is a quasi-random array of nodes (with given masses) in 3D connected by springs. It is formulated in small strain. The lattice nodes are connected by two springs, one representing the normal and the other shear contact stiffness. The springs represent elasticity of the rock mass. In *HF Simulator*, the calibration factors for spring stiffness are built-in and the user may specify typical macroscopic elastic properties as it is done for other conventional numerical models. The tensile and shear strengths of the springs control the macroscopic strength of the lattice. As for elastic constants, calibration factors are built-in for the strength parameters.

The model simulation is carried out by solving an equation of motions (three translations and three rotations) for all nodes in the model using an explicit numerical method. The following is the central difference equation for the translational degrees of freedom:

$$\dot{u}_i^{(t+\Delta t/2)} = \dot{u}_i^{(t-\Delta t/2)} + \Sigma F_i^{(t)} \Delta t / m$$
$$u_i^{(t+\Delta t)} = u_i^{(t)} + \dot{u}_i^{(t+\Delta t/2)} \Delta t$$

$$(1)$$

where $\dot{u}_i^{(t)}$ and $u_i^{(t)}$ are the velocity and position (respectively) of component i ($i=1, 3$) at time t, ΣF_i is the sum of all force-components i, acting on the node of mass m, with time step Δt. The relative displacements of the nodes are used to calculate the force change in the springs:

$$F^N \leftarrow F^N + \dot{u}^N k^N \Delta t$$
$$F_i^S \leftarrow F_i^S + \dot{u}_i^S k^S \Delta t$$

$$(2)$$

where "N" denotes "normal," "S" denotes shear, k is spring stiffness and F is the spring force. If the force exceeds the calibrated spring strength, the spring breaks and the microcrack is formed. In other words, if $F^N > F^{Nmax}$, then $F^N=0$, $F_i^S=0$, and a "fracture flag" is set.

2.3. Fluid flow

Fluid-flow model and hydro-mechanical coupling are essential parts of *HF Simulator*, as a code for simulation of hydraulic fracturing. The fluid flow occurs through the network of pipes that connect fluid elements, located at the centers of either broken springs or springs that represent pre-existing joints (i.e., springs intersected by the surfaces of pre-existing joints). (The code also can simulate the porous medium flow through unfractured blocks as a way to represent the leakoff. This capability is not discussed further in this paper.) The flow pipe network is dynamic and automatically updated by connecting newly formed microcracks to the existing flow network. The model uses the lubrication equation to approximate the flow within a fracture as a function of aperture. The flow rate along a pipe, from fluid node "A" to node "B," is calculated based on the following relation:

$$q = \beta k_r \frac{a^3}{12\mu}\left[p^A - p^B + \rho_w g\left(z^A - z^B\right)\right] \tag{3}$$

where a is hydraulic aperture, μ is viscosity of the fluid, p^A and p^B are fluid pressures at nodes "A" and "B", respectively, z^A and z^B are elevations of nodes "A" and "B", respectively, and ρ_w is fluid density. The relative permeability, k_r, is a function of saturation, s:

$$k_r = s^2(3 - 2s) \tag{4}$$

Clearly, when the pipe is saturated, $s = 1$ and the relative permeability is 1. The dimensionless number β is a calibration parameter, a function of resolution, used to match conductivity of a pipe network to the conductivity of a joint represented by parallel plates with aperture a. The calibrated relation between β and the resolution is built into the code.

2.4. Hydro-mechanical coupling

In *HF Simulator*, the mechanical and flow models are fully coupled.

1. Fracture permeability depends on aperture, or on the deformation of the solid model.
2. Fluid pressure affects both deformation and the strength of the solid model. The effective stress calculations are carried out.
3. The deformation of the solid model affects the fluid pressures. In particular, the code can predict changes in fluid pressure under undrained conditions.

A new coupling scheme, in which the relaxation parameter is proportional to $K_R a / R$, where K_R is rock bulk modulus and R is the lattice resolution, is implemented in *HF Simulator*, allowing larger explicit time steps and faster simulation times compared to conventional methods that use fluid bulk modulus as a relaxation parameter.

3. Verification test: Penny-shaped crack propagation in medium with zero toughness

The non-steady response of rock to injection of fluid depends on fracture toughness, the viscosity of the fluid and the rate of leak-off. In the case of zero fracture toughness and no leak-off, the response is viscosity-dominated, which corresponds to the "M-asymptote" identified by [5]. This condition is used for verification of *HF Simulator*.

In the simulated example, fluid is injected at a constant rate into a penny-shaped crack of low initial aperture (10^{-5}m). The crack has zero normal strength, and the in-situ stresses are also zero. Thus, the test conditions approximate those of the analytical solution for the no-lag case (i.e., no fluid pressure tension cut-off) provided by [5]. The injection rate is 0.01 m³/s; the dynamic viscosity is 0.001 Pa×s. The mechanical properties of the rock are characterized by Young's modulus of 7×10^{10}Pa and Poisson's ratio of 0.22. Figure 3 provides a visualization of the state of the model at 10 s of elapsed time. Note that pressures are negative in the outer annulus of the flow disk.

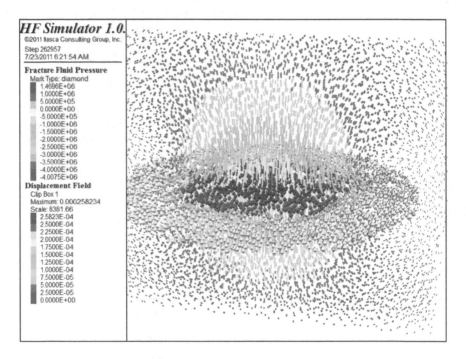

Figure 3. View of pressure (Pa) field (icons, colored according to magnitude) and cross-section of displacement (m) field (vectors, colored according to magnitude)

Figure 4 shows the aperture profiles at three times during the simulation — averaged numerical results (for 30 radial distances), together with asymptotic solutions (derived from the equations of [5]). Figure 5 shows the pressure profile at 10 s, together with the asymptotic solution. Note that there is a lack of match at small and large radial distances: at small distances, the numerical source is a finite volume, rather than a point source (which is assumed in the exact solution); at large distances, the finite initial aperture allows seepage (compared to zero seepage in the exact solution, which assumes zero initial aperture).

4. Example application

Two example problems are discussed in this section. Fracture propagation in a homogeneous (unfractured) and fractured media is analyzed. These two problems involve a horizontal borehole segment with two injection clusters with centers at 4.8 m distance (Figure 6). The model domain is 18 m × 18 m × 18 m, and the lattice resolution was set to 0.5 m. Fluid is injected into the clusters at rate of 0.01 m³/s. The assumed stress state is anisotropic with σ_{xx} = 1 MPa, σ_{yy} = 12 MPa and σ_{zz} = 10 MPa. The least principal stress is aligned with the horizontal section of the borehole. This stress state favors crack propagation in the direction normal to the horizontal section of the borehole. In order to initiate the fluid calculation, fluid-filled joints have been placed at the center of each cluster; these joints are slightly larger than the cluster size. The initial apertures in these joints have been set to 0.1 mm. Both example problems use this model configuration. The example shown on the left in Figure 6 simulates the response of an unfractured medium to fluid injection. Three discrete joints that interact with the induced fractures are introduced in the example on the right in Figure 6.

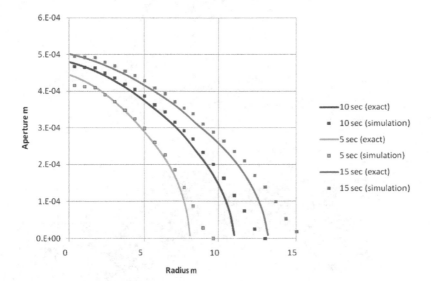

Figure 4. Aperture profiles for three times

The induced microcracks in the homogeneous model after 15 s of injection are shown in Figure 7. The microcracks form two roughly circular (penny-shape) hydraulic fractures. In this example, the fractures are not parallel. There is a slight trend of fractures curving away from each other as a result of stress interaction.

In the second example, the HF propagation is clearly affected by the pre-existing joints, as shown in Figure 8. When the HF intersects the pre-existing joint, the fluid is diverted into the pre-existing joints. (In general case, the HF can cross or be diverted into the pre-existing joint, depending on a number of parameters, including stress state, strength and permeability of the pre-existing joint.) The propagation continues by reinitiation along the edges of pre-existing joints.

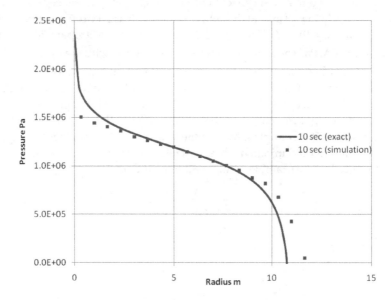

Figure 5. Pressure profile at 10 s

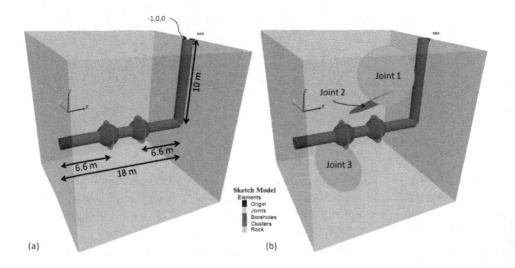

Figure 6. Geometry of two example problems

5. Conclusion

HF Simulator is a powerful 3D simulator for hydraulic fracturing in jointed rock mass that allows the main mechanisms (nonlinear mechanical response, fluid flow in joints and coupled fluid-mechanical interaction) to be reproduced. The formulation of *HF Simulator* is based on a quasi-random lattice of nodes and springs.

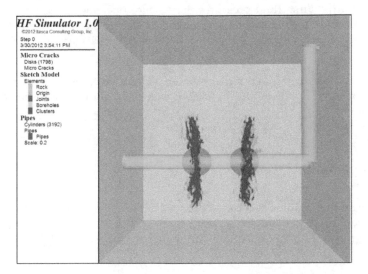

Figure 7. Hydraulic fractures generated in a homogeneous medium (dark blue disks are microcracks)

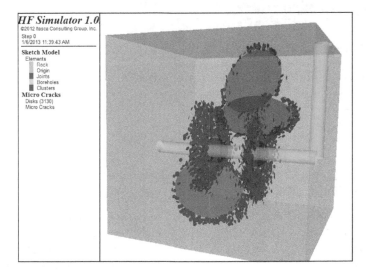

Figure 8. Hydraulic fractures generated in a medium with three pre-existing joints (blue disks are microcracks)

The springs between the nodes break when their strength (in tension) is exceeded. Breaking of the springs corresponds to the formation of microcracks, and microcracks may link to form macrofractures. The SJM (smooth joint model) is used to represent pre-existing joints in the model. Thus, the SJM allows simulation of sliding of a pre-existing joint in the model, unaffected by the apparent surface roughness resulting from lattice resolution and random arrangement of lattice nodes.

The model is fully coupled hydro-mechanically. There are several ways in which fluid interacts with the rock matrix. First, fluid pressures may induce opening or sliding of the fractures. Second, mechanical deformation of fractures causes changes in joint pressures. Third, the mechanical deformation changes the permeability of the rock mass as the joint apertures change.

The new code is a promising tool for simulation and understanding of complex processes, including propagation of HF and its interaction with DFN, during stimulation of unconventional reservoirs.

Acknowledgements

The development of the numerical code described in this paper was funded by BP America. The authors would like to thank BP America for their support. Matt Purvance, Jim Hazzard and Maurilio Torres of Itasca Consulting Group, Inc. are thanked for their valuable work on HF Simulator.

Author details

B. Damjanac, C. Detournay, P.A. Cundall and Varun

Itasca Consulting Group, Inc., Minneapolis, Minnesota, USA

References

[1] Potyondy, D. O, & Cundall, P. A. A Bonded-Particle Model of Rock. Int. J. Rock Mech. & Min. Sci., (2004). , 41, 1329-1364.

[2] Pierce, M. Mas Ivars D., Cundall P.A., Potyondy D.O. "A Synthetic Rock Mass Model for Jointed Rock," in Rock Mechanics: Meeting Society's Challenges and Demands (1st Canada-U.S. Rock Mechanics Symposium, Vancouver, May 2007), Fundamentals, New Technologies & New Ideas, E. Eberhardt et al., Eds. London: Taylor & Francis Group; (2007). , 1, 341-349.

[3] Itasca Consulting GroupInc. PFC2D (Particle Flow Code in 2 Dimensions), Version 4.0. Minneapolis: Itasca; (2008).

[4] Itasca Consulting GroupInc. PFC3D (Particle Flow Code in 3 Dimensions), Version 4.0. Minneapolis: Itasca; (2008).

[5] Peirce, A, & Detournay, E. An Implicit Set Method for Modeling Hydraulically Driven Fractures, Comput. Methods Appl. Mech. Engrg., (2008). , 197, 2858-2885.

Mining and Measurement

Monitoring and Measuring Hydraulic Fracturing Growth During Preconditioning of a Roof Rock over a Coal Longwall Panel

R. G. Jeffrey, Z. Chen, K. W. Mills and S. Pegg

Additional information is available at the end of the chapter

Abstract

Narrabri Coal Operations is longwall mining coal directly below a 15 to 20 m thick conglomerate sequence expected to be capable of producing a windblast upon first caving at longwall startup and producing periodic weighting during regular mining. Site characterisation and field trials were undertaken to evaluate hydraulic fracturing as a method to precondition the conglomerate strata sufficiently to promote normal caving behaviour at longwall startup and reduce the severity of periodic weighting. This paper presents the results of the trials and illustrates the effectiveness of hydraulic fracturing as a preconditioning technique.

Initial work was directed at determining if hydraulic fractures were able to be grown with a horizontal orientation, which would allow efficient preconditioning of the rock mass by placing a number of fractures at different depths through the conglomerate from vertical boreholes drilled from the surface. The measurements and trials were designed to determine the in situ principal stresses, the hydraulic fracture orientation and growth rate, and whether the fractures could be extended as essentially parallel fractures to a radius of at least 30 m. Overcore stress measurements were used to determine the orientation and magnitude of the in situ principal stresses, a surface tiltmeter array was used to determine the hydraulic fracture orientation, and stress change monitoring, pressure monitoring and temperature logging in offset boreholes were used to establish the fracture growth rate, lateral extent, and that the fractures maintained their initial spacing to a radial distance of greater than 30 metres. The measurements and trials demonstrated that horizontal fractures could be extended parallel to one another to a distance of 30 to 50 m by injection of 5,000 to 15,000 litres of water at a rate of 400 to 500 L/min. Results from the trial allowed a preconditioning plan to be developed and successfully implemented.

1. Introduction

Hydraulic fracturing has been applied successfully to preconditioning of hard rock at several block caving mines [1-3] and has been used to weaken a sandstone channel over a longwall panel [4]. A recent paper documents related work in China applied to control of rock bursting [5]. In addition, hydraulic fracturing has been used to induce caving in block caving operations [1] and in longwall coal mining [6]. The work described in this paper applied hydraulic fracturing to preconditioning of a strong roof rock in order to weaken it to promote earlier caving during start up of a longwall.

Narrabri Coal Operations, located 28 km south of Narrabri, NSW, are extracting the Hoskissons coal seam using a 300 m wide longwall. The Digby conglomerate is 15 to 20 m thick and lies immediately above the seam. Geotechnical assessments of its potential to cave during longwall mining concluded that this conglomerate would not cave into the goaf until more than 60 m of the coal was extracted [7]. In addition, the analysis highlighted the potential for the conglomerate to pose a periodic weighting risk.

Periodic weighting occurs when the roof strata is strong enough to support itself behind the longwall face for some distance before failing suddenly as mining progresses. Failure typically occurs just ahead of the face and may cause the longwall supports to become overloaded and converge, crushing the coal on the face and posing a rock fall hazard for equipment and miners located between the face and the supports. The project described in this paper was aimed to test hydraulic fracturing as a method to precondition the conglomerate sequence and promote caving during mining.

The preconditioning test program involved initial characterisation of the in situ stresses to determine the suitability of the site for placing hydraulic fractures with a horizontal orientation. The stress measurement work was followed by a three stage program of field trials. The first stage was aimed to confirm that hydraulic fractures were able to be formed horizontally and extended for a distance of more than 30 m from the injection hole, given the site conditions and the available equipment. The second stage was aimed to confirm that multiple hydraulic fractures placed in close vertical proximity remained essentially parallel to each other. The third stage was aimed to confirm conditions remained suitable to form horizontal fractures in a deeper area of the mine.

The field trials used an array of monitoring boreholes drilled at various distances around a central injection hole. During the first stage of the trials, five fractures were placed through the conglomerate sequence using a straddle packer system. These fractures, which were placed at a depth of 140 to 160 m, were monitored by a surface tiltmeter array, by boreholes offset 10 to 30 m from the borehole being fractured, and by stress change monitoring instruments located at 25 m and 60 m from the injection hole. Acoustic image logs of the injection hole and boreholes intersected by hydraulic fractures and core from intersected boreholes provided additional confirmation that fractures were able to be formed horizontally.

For the second stage, a second injection hole was drilled offset from the first borehole. The bottom hole locations of these two boreholes, A and J, were determined by survey to be separated by 6.5 m. Five fractures were placed with a vertical spacing of 2.5 m and each of these was later reopened to determine intersection depths in borehole C and growth rate to boreholes A and E (Figure 1). As well as monitoring used in the first stage, temperature logging of borehole interesections in borehole C confirmed that multiple fractures were able to be formed parallel to each other over an extended distance.

The third stage of the program was conducted at the start of the third longwall panel in an area where the overburden depth is some 20-30 m deeper than at the first site. A single injection hole and two monitoring holes confirmed that hydraulic fractures were able to be formed horizontally at this location despite the greater overburden depth.

This data set provides evidence for hydraulic fracture growth to more than 30 m radius at a vertical spacing between fractures of 1.25 m and 2.5 m, with non-symmetric fracture growth measured by the offset borehole data.

1.1. Design approach

Two sites were instrumented and tests were carried out to verify hydraulic fracture growth behaviour and measure the parameters needed to design the hydraulic fracture preconditioning process. Figure 1a shows the two test sites and their relative location with respect to each other and to the longwall panels at the mine. Figures 1c and 1d contain scale drawings of the sites, with the fracturing and monitoring boreholes indicated. Both sites had a surface tiltmeter array installed and the tiltmeter instrument locations at the sites are indicated in the figures.

The second site was located over the start of Longwall 103 where the conglomerate lies at a depth of 162 to 181 m (see Figure 1d). The fractures at the Longwall 103 site were placed into borehole 103AA with temperature logging occurring in monitor boreholes 103ACRR and 103AB. The temperature logging served to detect the arrival of the fractures at these boreholes and to locate their vertical positions in the boreholes.

The fractures at the Longwall 103 site were placed into borehole 103AA with temperature logging occurring in monitor boreholes 103ACRR and 103AB. The temperature logging served to detect the arrival of the fractures at these boreholes and to locate their vertical positions in the boreholes.

2. Preconditioning plan

By placing a number of horizontal fractures through the conglomerate layer, the mechanical behaviour of the conglomerate is modified from a thick-plate behaviour to a series of much thinner stacked plates with the aim of promoting caving. For efficient preconditioning from vertical holes drilled from the surface, hydraulic fractures are required to form horizontally as shown in Figure 1b. This fracture orientation allows efficient preconditioning from a vertical borehole because multiple fractures can be placed from each borehole.

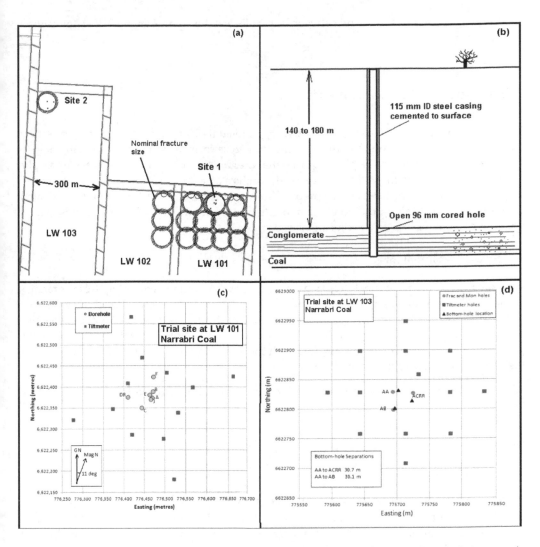

Figure 1. Test sites and borehole layout used. The drawings show (a) the location of the two test sites relative to each other and the planned underlying longwall roadways, (b) a vertical section showing the preconditioning concept, (c) the borehole layout at test site 1 over LW101 with tiltmeter locations, and (d) the borehole layout at test site 2 over LW103 with tiltmeter locations.

To confirm that this strategy would be possible in the local site conditions at Narrabri, a trial was conducted to determine the fracture orientation, growth rate, and to verify that the fractures could be extended parallel to previous fractures for a distance of 30 m or more. For example, boreholes spaced at 80 m centres require fractures to grow to 45 m radius with each borehole preconditioning 6,300 m² of conglomerate and a 50 m spacing between holes would require fractures to grow to 30 m with each borehole then preconditioning 2,800 m² of the conglomerate. These parameters would then be used to establish that horizontal fractures were feasible, to determine the spacing between boreholes and the volume and rate of water to inject per fracture.

3. Test sites and measurements

For the preconditioning to work efficiently, the minimum principal stress must be the vertical stress. This allows horizontal hydraulic fractures to be formed. In addition to being able to grow horizontal fractures, the plan required that the growth rate and ultimate size of hydraulic fractures be determined so that the vertical borehole spacing could be specified. A second important requirement for effective preconditioning was that the fractures could be initiated at regular intervals along the borehole and extended as essentially parallel fractures, with a spacing of 2.5 m or less, to a radius of 30 m or more.

3.1. Stress measurements

As a starting point, the vertical stress was estimated by integrating density logs from boreholes drilled in the same area. Two overcore stress measurements from a vertical borehole were then made at 145 m depth in the conglomerate using ANZI strain cells [8]. Each ANZI cell contains 18 strain gauges which, when bonded to the surface of a pilot hole, sense the rock strain as the gauge is overcored. Using this strain data with the rock modulus, measured in an independent test, the in situ stress acting can be found.

Analysis of this data gave an estimate of the principal stresses acting as [9]:

σ_1= 8.3 MPa ± 1.1,

σ_2= 4.7 MPa ± 0.9, and

σ_3= 4 MPa ± 2.4,

with σ_3 the vertical stress and σ_1 directed N30E. The accuracy of the vertical stress in these tests is believed to have been affected by proximity to a small geological fault structure that was not recognised until later when the underground roadways were developed. The plus and minus range listed above for each stress component represents two standard deviations. For comparison, integration of a density log from a vertical borehole gave an estimate of 3.1 MPa for the vertical stress at 145 m depth. Earlier biaxial overcore stress measurements [10] gave generally horizontal total stresses of σ_1= 7.7 and σ_2= 6.0 MPa in the conglomerate in borehole NC-098 at a depth of 156.7 m with both values well above the log based vertical stress magnitude of 3.35 MPa for this depth. The value of the vertical stress from the density log was taken as an accurate value because bedding anisotropy can affect the vertical stress measured by the ANZI cell. In addition, the instantaneous shut-in pressure and offset monitor pressure data were found to correlate well with the log derived vertical stress magnitude. Taken together, the stress measurements gave an indication that horizontal fractures were likely to form, but because σ_3 and σ_2 as measured by the ANZI cell were of similar magnitude, this inference needed to be verified by placing full scale hydraulic fractures, monitored using tiltmeters and offset boreholes.

3.2. Fracture asymmetry measurement

As a series of hydraulic fractures are placed sequentually into a borehole, with the fractures placed one above the other, there is potential for them to interact. During preconditioning, the hydraulic fractures are placed at a rate of approximately one per hour. The fractures are not propped, but some injected fluid remains in the fracture and bleeds back into the well once the packers are moved uphole in preparation for the next treatment. These fractures induce a change in the stress field around them and this changed stress will affect the next fracture, potentially causing it to curve toward or away from the previous fractures and to grow asymmetrically. Figure 2 shows the stress changes measured by an ANZI strain cell located at 129.3 m below the surface during the placement of the first hydraulic fracture in borehole A. The peak stress observed approximately 17 m above the hydraulic fracture was 0.52 MPa soon after the hydraulic fracture was placed at a depth of 146.5 m and this stress had reduced only to 0.26 MPa some 1.5 hours later at which time the excess pressure in the fracture was 0.3 MPa. Once one fracture grows somewhat asymmetrically, the next fracture is likely to find it easier to grow in a way such that it grows so as to avoid the residual vertical stresses created by the previous fracture and its centre of volume is offset relative to the centre of volume of the previous fracture.

The movement of the fracture centre of volume can, in principle, be detected by analysis of the tilt data [11] and also by noting the time of intersection of the fracture with the monitoring boreholes. Both of these methods of detecting asymmetry were used for the fracturing work carried out at the test sites.

Figures 3 and 4 summarise data recorded during fracture 4J and 7J, showing both the injection pressure at borehole J and the pressure response in the monitored boreholes. During fracture 4J, single packers were installed at the top of the conglomerate, set at 140.9 m to the bottom of the packer rubber, in monitor boreholes C and E and a vibrating wire piezometer was located at 146.0 m in monitoring borehole A. The piezometer was intalled in a coarse sand-filled section of the borehole with a grout plug at the conglomerate base and a second grout plug placed from the top of the conglomerate to the surface. The packers each contained a mandrel that connected through the packer to the open hole below. This pressure was transmitted to the surface via a 6 mm ID high pressure hose which was connected to a pressure transducer for logging. The pressures shown for the injection pressure and for boreholes C and E have had the hydrostatic pressure to the depth of the injection point in borehole J added to them to give an approximate bottom hole pressure. The calibrated piezometer output gives a direct bottom hole pressure at its set depth in borehole A.

Fracture 4J was carried out by straddling a slot at 151.8 m in borehole J. The fracture grew into boreholes E, C and then A as indicated by the pressure responses shown in Figure 3. In order to fit a circular fracture to this implied growth, the centre of the fracture needs to be located,

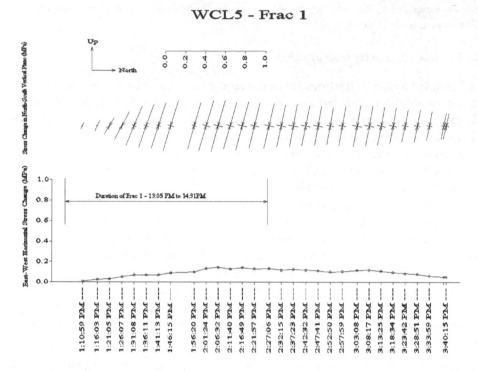

Figure 2. Stress change recorded during fracturing of borehole A. The fracture was shut-in at 14:31 (2:31 pm) and flowed back at 17:20 (5:20 pm) so pressurised fluid was contained in the fracture during the entire period of this plot. Fracture closure occured at 16:08 (4:08 pm).

at the time the fracture grows through borehole A, at a point 15 m west of borehole J. The intersection The intersection time for borehole A corresponds to a pressure at the piezometer of 3.47 MPa, which is just above the vertical stress magnitude. However, the earlier rise in pressure at 9:56 could be an indication of an earlier intersection, although the pressure at that time only reaches a value of 2.56 MPa. If water was being lost out of borehole A above the fracture depth (near 151.8 m), perhaps into an existing hydraulic fracture connected into borehole J above the packers, then this flow through the coarse sand would make the pressure in borehole A non-uniform and would result in the piezometer reading a pressure lower than the pressure in the fracture located approximately 5.8 m below it. However, the coarse sand used has an estimated permeability of 2,000 Darcies. A flow of 17 L/min through 5 m of this sand pack would result in a steady-state pressure drop of only 0.1 MPa. An earlier intersection time would support a less asymmetric fracture shape development and does highlight a possible source of error in picking intersection times based on pressure measured at borehole A.

Data collected during fracture 7J is summarised in Figure 4. The intersection with boreholes E and A occurred close together in time. In this case, the fracture depth (146.8 m) is very nearly the same as the piezometer depth (146.0 m) in borehole A, which minimises the issue of water flowing through the sand pack affecting the piezometer pressure. During this fracture treatment, temperature logging was carried out in borehole C, which was open at the borehole collar. The intersection time of the fracture into borehole C is indicated in Figure 4 and the temperature log is shown in Figure 5. The temperature logging method involved first cooling the water in the borehole by pumping ice water through a 20 mm diameter polypipe to the bottom of the borehole. A cooled condition of 10°C or less was typically achieved. The rock temperature at 145 m is approximately 23°C at this site and the water injected into a hydraulic fracture is quickly warmed to this temperature. Therefore, intersection locations were found by noting the depth where warm fluid was entering the monitored borehole and the first arrival of warm fluid into the hole is an indication of fracture growth rate. The sensor located at 158.5 m in borehole C started to increase in temperature at 11:50 (see Figure 5 and by 11:57 two warm peaks had been established at 145.5 m and 147.9 m. Early and weaker warming events may be associated with fluid being expelled from previously placed hydraulic fractures which are squeezed more tightly shut as the propagating hydraulic fracture interacts with them. The stronger warming events at 11:56:08 are therefore taken as corresponding to the intersection time.

Intersections from a number of fractures placed into borehole J have been used to define the fracture growth asymmetry. Figure 6 shows the range of fracture asymmetry measured by intersection data from this analysis. Only circular fractures are considered in Figure 6, although it is believed unlikely that the fractures were perfectly circular. However, if the fractures are allowed to take non-circular shapes, the range of centre locations and fracture sizes that can be fitted to the intersection data is increased significantly.

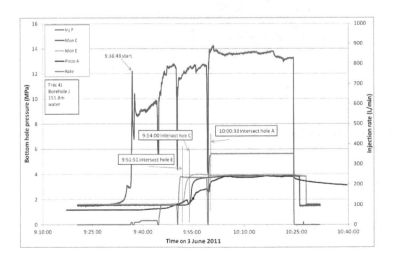

Figure 3. Data summary plot for fracture 4J, which includes pressure monitoring data in boreholes A, E, and C.

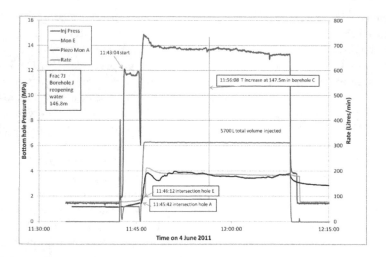

Figure 4. Data summary plot for fracture 7J, including pressure monitoring data in boreholes A and E. Temperature was monitored in borehole C during this fracture (see Figure 5).

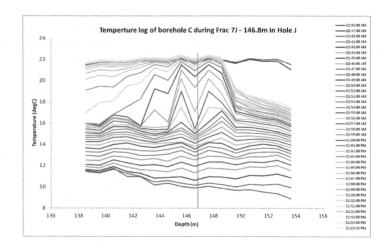

Figure 5. Temperature variation with depth in borehole C during fracture 7J. A line has been drawn at 146.8 m to indicate the nominal depth of fracture 7J in hole J.

3.2.1. Asymmetry from analysis of tilt data

By representing the fracture in the forward model as a displacement discontinuity (DD) singularity within a homogeneous, isotropic linear-elastic half-space and using a Bayesian probabilistic inversion approach, the fracture volume, orientation, and location of the fracture center-of-opening versus time have been estimated by analyzing the tilt measurements. This provides considerable insight into the geometry and development of the hydraulic fractures.

The displacement field produced by a DD of the intensity D_j across a surface S (the hydraulic fracture) in a uniform elastic half space can be expressed as [12]

$$u_i(x) = \iint_S D_j(x') \left[\delta_{jk} \lambda \frac{\partial U_{il}}{\partial x_l} + \mu \left(\frac{\partial U_{ij}}{\partial x_k} + \frac{\partial U_{ik}}{\partial x_j} \right) v_k \right] ds \tag{1}$$

where $u_i(x)$ is the displacement in the x_i direction at a point x. $U_{ij}(x, x')$ is the ith component of displacement at x due to a point force of unit magnitude acting in the x_j direction at a point x' on S within an elastic half space. v is the normal to S at point x'. λ and μ are the Lame coefficients for the elastic rock material.

The measured tilt angles are related to the displacement gradients by

$$\omega_1 = \frac{\partial u_1}{\partial x_3} - \frac{\partial u_3}{\partial x_1}, \omega_2 = \frac{\partial u_2}{\partial x_3} - \frac{\partial u_3}{\partial x_2} \tag{2}$$

For a horizontal hydraulic fracture which grows symmetrically with respect to the borehole, the fracture centre is taken to be at the injection point. Sometimes, asymmetric growth of the hydraulic fracture can occur. In this case, the fracture centre will move away from the injection point as the fracture grows. It is assumed that the fracture is planar, so the injection point and the fracture centre must remain in the fracture plane.

It has been shown that in most cases the analysis of tilt data allows for a robust estimation of fracture volume and orientation (dip and strike) [13, 11]. To investigate the movement of the fracture centre, consider a DD singularity centered at $(x_c, 0,0)$ in an infinite elastic body. Given an offset, which for simplicity we specify as along only the x axis, of $(\Delta x_c, 0,0)$ for the DD center, the tilt component can be obtained by using a Taylor series expansion as

$$\omega_1^{DD}(x_c + \Delta x_c) = \omega_1^{DD}(x_c) \left[1 + \left(\frac{\Delta x_c}{r} \right) g_1 + \left(\frac{\Delta x_c}{r} \right)^2 g_2 + O\left(\frac{\Delta x_c}{r} \right)^3 \right] \tag{3}$$

where $r = \sqrt{(x - x_c)^2 + y^2 + z^2}$ denotes the distance between the point (x, y, z) and the DD center $(x_c, 0,0)$, and the functions g_1 and g_2 are of order $O(1)$.

Eq. 3 shows that the estimation of fracture center movement is coupled with the tilt measured at x_c which depends on the fracture orientation and volume. The fracture center movement Δx_c is difficult to be resolved when it is far less than the observation distance r.

Two synthetic examples are presented here to show the effect of fracture center movement on the estimation of fracture volume and orientation. In the first example, the synthetic tilt data are generated by using a point DD singularity with a dip of 20^0 and a dip orientation of N160^0 in an elastic half-space. The fracture center is fixed at 20 m east of the injection point. The fracture volume increases linearly with time, reaching a maximum of 6 m^3 at 40 minutes. Then the generated tilt data are used to infer the fracture geometry by using a Bayesian probablistic inverse approach, assuming that the fracture is centered at the injection point. The predicted fracture dip direction and dip are shown in Figure 7. As we can see, an incorrect assumption on the fracture center location leads to a poor prediction of the fracture orientation.

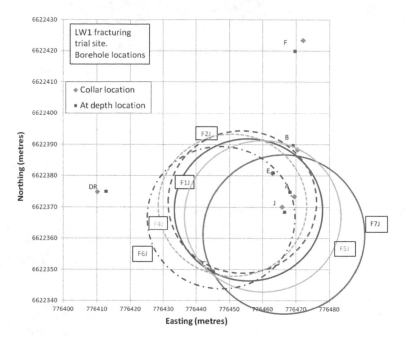

Figure 6. Longwall 101 trial site showing approximate fracture asymmetry inferred from intersection data for fractures placed into borehole J. Fractures are drawn at the time of the last intersection and in most cases injection stopped shortly after this time.

In the second example, the fracture has a dip of 10^0 and a dip orientation of $N160^0$ centered at the injection point. The synthetic tilt data are used to infer the fracture geometry (fracture volume and dip orientation) and the fracture center movement (see Figure 8) by specifying the frature dip of 20^0. It can be seen from Figure 8 that the incorrect constraint of the fracture dip results in an incorrect inferred movement of the fracture center.

Table 1 contains the fracture center location inferred from analysis of the tiltmeter data for a number of fractures in borehole J. Because the location of the center of volume is correlated to the dip and dip direction, the analysis was carried out for a case where both the orientation and the center of volume were found with no constraint and then again for the case where the fracture was constrained to be horizontal.

Of the fractures listed in Table 1 and also drawn in Figure 6 with their locations based on intersection data, only fracture 7J has an inferred center of volume that is somewhat consistent for the two methods. The tiltmeter results, which correspond to a time of 15 minutes from the start of injection, generally indicated less movement of the fracture center than the intersection data suggests. The inferred dip magnitude from the tiltmeter analysis is in the range of 40^0 for fractures 1J, 4J, and 6J. This dip magnitude is larger by a factor of at least two than dips inferred from tilt data for fracturing carried out in other boreholes in this area (see Table 2, average dip 15.9⁰) and does not agree with the intersection data either. The reason for these relatively large dips for this series of fractures, which are thought to be in error, is not know but may be related to small movements induced on faults present in this portion of the longwall.

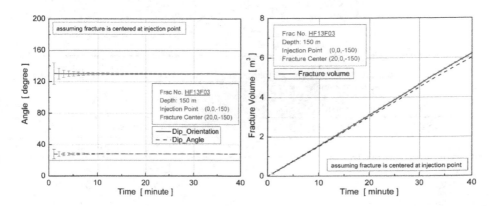

Figure 7. The predicted fracture geometry calculated with an assumed fracture center located at the injection point when in fact it is offset by 20 m.

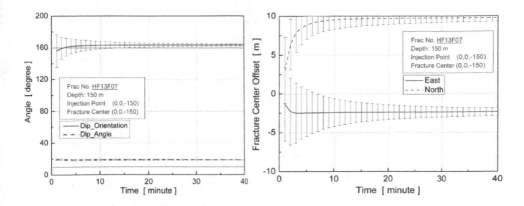

Figure 8. The predicted fracture geometry and center movement obtained by specifying the fracture has a dip of 20⁰ when in fact the dip is 10⁰.

Fracture	Northing offset (meters)	Easting offset (meters)	Dip/Dip Direction (degrees)
1J	0	0	C_0
1J	+0.5	0	38/180
4J	0	0	C_0
4J	0	0	38/160
6J	-1.0	0	C_0
6J	+4.0	+2.0	40/160
7J	-5.0	+1.0	C_0
7J	-1.0	+2.0	25/170

Table 1. Centre location of fractures relative to borehole location based on tiltmeter data. Fractures with a dip of C_0 were constrained to be perfectly horizontal.

Borehole	Fracture	Orientation	
		Dip	Dip direction
A	2	20^0	185^0
	4	10^0	180^0
101AW	1	22^0	330^0
	5	2^0	140^0
101BAR	1	20^0	70^0
	3	2^0	5^0
101ASR	1	30^0	190^0
	7	20^0	185^0
	10	25^0	190^0
102AA	1	20^0	300^0
101AUR	2	35^0	140^0
	3	15^0	165^0
102AE	2	10^0	65^0
102AD	1	10^0	190^0
101AL	1	8^0	270^0
	3	5^0	300^0

Table 2. Variation of dip and dip direction as determined by tiltmeter analysis.

For comparison, the dip direction and dip for fracture 3 at the Longwall 103 test site is shown in Figure 9a with the tilt vertors from this fracture at the end of the injection shown in Figure 9b. The 20° dip is believed to be too large and is likely to be reflecting some movement of the centre of opening of the fracture.

3.3. Fracture growth measurement

Tiltmeter monitoring [11-15], stress change monitoring [16], offset borehole measurements [17], and fracture growth modelling using a numerical hydraulic fracturing model were used to obtain the fracture growth rate as a function of injection rate and volume. The tiltmeter data provided a confirmaton that the hydraulic fractures were essentially horizontal in orientation. Stress change monitoring using ANZI cells installed in boreholes B and F (Figure 1), indicated fracture growth below these locations during injections into borehole A.

The primary data used to establish the hydraulic fracture growth rate were timing of the first arrival intersection events at offset boreholes. These data were filtered to remove the most extreme asymmetric growth cases so that an axisymmetric hydraulic fracture model could be used to match the measured growth. By matching several different measurements, the model was calibrated for the conditions at the Narrabri site. The calibrated model was then used to produce a set of time versus fracture radius curves for three different injection rates and these were then used for choosing a rate and volume that would produce a fracture size needed in the preconditioning work. A borehole spacing compatable with the ultimate size of the fractures was selected as part of this process. Figure 10 contains the growth curves generated by the numerical fracture model with several points indicating measured intersection events, for fractures placed using similar rates, also shown.

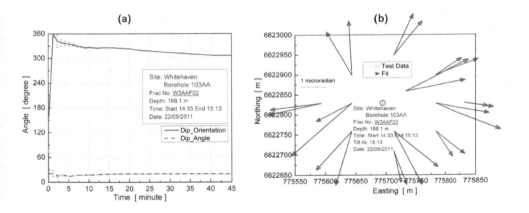

Figure 9. Analysis of tilt data from fracture 3 at the LW3 site. The fracture was interpreted to dip to the north at 20°. The tilt vectors at the end of the injection are consistent with a horizontal fracture deformation field.

The curves calculated were fitted to the higher growth rate data represented by Frac 2 in borehole 101AM. In this case, the fracture grew through two monitoring holes located 30 m to the east and 30 m to the west of 101AM. The growth for this fracture was therefore thought to be fairly symmetric. The other measurements shown are from the trial site over Longwall 103 where injection occured into borehole 103AA and monitoring occurred at two offset boreholes. These points illustrate the variability in the measured growth data with asymmetric growth being a primary cause. For example, Frac 3 in 103AA grew through one monitoring borehole located 31 m to the east of the injection borehole after 11 minutes, but required 23 minutes to grow through the second monitor hole located 30 m to the south of the injection borehole. On average, growth of the fractures seemed to be somewhat slower at the Longwall 103 site than at the Longwall 101 sites. Using these data, the treatments for preconditioning of the main longwall panels were designed to inject water at 500 L/min for 25 minutes each which, according to the growth curves in Figure 10, would produce fractures of approximately 45 m radius. The boreholes over the main Longwall 101 panel were drilled using an 80 m spacing.

3.4. Fracture vertical spacing measurement

To achieve the intended degree of treatment of the conglomerate, it was desirable to create hydraulic fractures that were parallel to one another so that the massive conglomerate layer was divided by the fractures into thinner and mechanically weaker system of layers. Work by

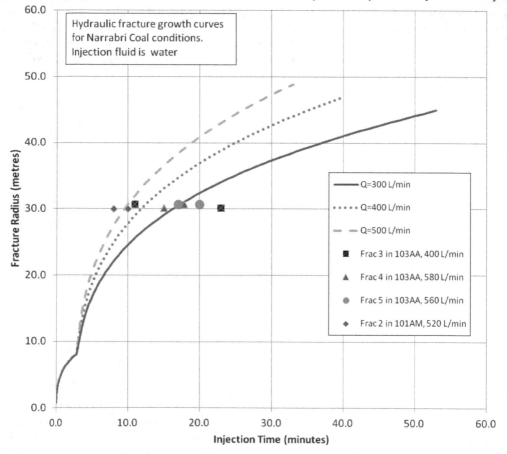

Figure 10. Growth measurements with curves generated by a numerical fracturing model.

[3] has shown that fractures grown through rock blocks in the laboratory are sensitive to initiation conditions. This laboratory work showed that fractures that initiated by splitting the borehole and then reoriented to become transverse to the borehole, were found to have much less regular spacing away from the borehole with growth into adjacent fractures commonly occurring (see [18] for additional details of the laboratory determined fracture paths). In contrast to this, if the fractures were initiated from slots or notches cut into the borehole, the fractures initiated and continued to grow in the same plane, as transverse fractures. The laboratory work showed that these fractures were spaced more regularly and they tended to extend into the far-field as parallel fractures. The initiation sites in the vertical boreholes at Narrabri were therefore notched using a sand and water abrasive jetting method. By rotating the jetting tool slowly, a circumferential slot was cut to act as a stress concentration point for fracture initiation.

A separate laboratory study, described in [19], presents results verifying a theory of closely spaced hydraulic fracture growth. The theory applies to fractures that are placed successively, one after the other and can be applied to predicting if the next fracture in a sequence will curve towards or away from the previous fracture. To make a prediction, the size and residual width of the previous fracture must be known. In addition, the rock elastic properties, the coefficient of friction for sliding of the hydraulic fracture surfaces, the stress field, and the injection rate and fluid viscosity are required. These parameters are then inserted into expressions for several dimensionless groups whose value then determine the type of curving to expect (see [19] for details). The calculation applied using parameters for the conditions at Narrabri, predict that essentially no curving of the hydraulic fracture will occur and that the fractures should grow parallel to one another.

Intersection of the fractures with offset pressure monitoring boreholes, as shown in the data contained in Figures 3 and 4, confirmed an approximately horizontal fracture orientation, but did not confirm that the fractures were growing parallel to one another. Fractures may have been growing into an adjacent fracture, for example, or growing with divergent paths which would leave wedge-shaped block of unfractured rock. The acoustic image logs run before and after fracturing were not able to detect the horizontal fractures in the horizontally bedded conglomerate. Two fracture traces, both dipping at approximately 10° to the west are visible at 149.9 and 150.1 m in the acoustic image of borehole J before it was fractured. These may be hydraulic fractures generated during fracturing of borehole A, but no other fracture traces can be seen in this image suggesting the hydraulic fractures are not wide enough to be seen by this method.

The core from borehole J, which was drilled after borehole A was hydraulically fractured, was examined in order to detect the fractures placed into borehole A. Several horizontal fractures were logged with four of them showing rotational shearing caused by the core rotating at that point during drilling. Such rotationally sheared fractures are normally fairly rare and have, in this case, been taken as indications of the location of the hydraulic fractures. Table 3 compares the fracture depths in borehole A with the depths of the logged rotationally sheared fractures in borehole J. The rotated core breaks in the core from borehole J are found to correspond

Fracture	Depth of initiation (metres)	Depth from core (metres)
1	146.4	146.1
2	149.8	150.0
3	140.2	140.2
4	144.7	
5	151.5	151.9

Table 3. Comparison of fracture initiation depths in borehole A with rotated core breaks from borehole J.

closely to the depths of initiation of the fractures in borehole A, suggesting nearly horizontal fracture orientation. This conclusion is supported by other intersection data such as that shown in Figure 4 for fracture 7J, which grew from borehole J into borehole C, 28.8 m away, and was located by termperature logging to be within plus or minus 1 m of the initiation depth (an apparent dip of 2°). The horizontal distances between the injection boreholes and the monitoring boreholes are listed in Table 4.

Borehole	A	J
A	0.0 m	6.5 m
B	14.9 m	21.3 m
C	34.4 m	28.8 m
DR	55.3 m	54.1 m
E	8.0 m	13.0 m
F	45.1 m	51.4 m
J	6.5 m	0.0 m

Table 4. Horizontal distance in metres between boreholes A and J and other boreholes at the site.

Inversion of the tiltmeter data cosistently produced dips of 10° to 30°. These larger dips seem to be in error in light of the nearly horizontal orientations obtained from intersection data with temperature logging. Figure 11 shows the fracture spacing implied by the temperature logging carried out in borehole C during fracturing of borehole J. Figure 12 shows fracture vertical spacing measured while fracture borehole 101AM while temperature logging in borehole 101AN. Both data sets illustrate the essentially parallel growth of the hydraulic fractures between these holes, confirming the prediction made from the theory of closely spaced fracture growth. Holes AM and AN lie along the startup roadway running at an azimuth of 273° with respect to grid north while the line connecting borehole J to C is oriented at an azimuth of 231°. Borehole AM is approximately 85 m NW of borehole J. Therefore, if these sections are taken as representative, the hydraulic fractures are essentially horizontal and maintain their initiation vertical spacing over more than 30 m of growth.

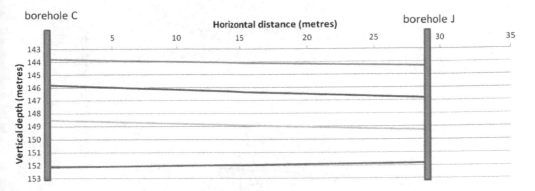

Figure 11. Fracture orientation and spacing implied by intersection and temperature logging data collected in borehole C during fracturing of borehole J. Fractures were placed at 2.5 m vertical spacing in borehole J.

4. Caving behaviour

The longwall started retreating on 12 June 2012 with a windblast management plan in place that required additional precautions to be used during mining until the caving commenced and the goaf developed. If the goaf behind the longwall face had not formed by the time the face had retreated to 25 m from the start position, additional work to induce caving was planned. However, the conglomerate caved, starting at the centre of the panel and progressing toward both gate roadways, after 24 m of retreat. This was a significant improvement over the estimated distance of more than 60 m for caving to start that was made based on modeling studies of the untreated conglomerate.

Beyond the startup area for a distance of 200 m, the conglomerate was preconditioned using boreholes located on approximate 80 m centres. The intensity of fractures placed in this main part of the longwall panel was approximately 25 percent of that applied along the startup section. A 100 m wide window was then left with no preconditioning to allow comparison of the fractured and unfractured conglomerate caving behavior. Mining under this section of conglomerate demonstrated that the preconditioning reduced the intensity of the periodic weighting events, but the events that still occurred were more random under the preconditioned roof. When mining under the conglomerate that was not preconditioned, weighting events could be anticipated to occur at regular intervals of about 15 m of longwall retreat. Therefore, adjustments to the daily longwall extraction plans were made so that any slowing or halting of mining was avoided when approaching an anticipated weighting event. Using this modified mining strategy, mining was continued without using preconditioning for the rest of Longwall 101.

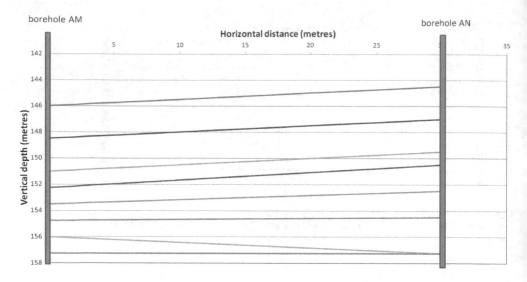

Figure 12. Fracture orientation and spacing implied by intersection and temperature logging data collected in borehole 101AN during fracturing of borehole 101AM. The upper part of AM was fractured at 2.5 m vertical spacing which was reduced to 1.25 m spacing below 151 m depth.

5. Conclusions

Measurements of fracture growth, spacing and orientation at two trial sites and as the preconditioning of the Longwall 101 startup area was carried out demonstrated that the hydraulic fractures could be created that were essentially horizontal and could be extended to more than 30 m as parallel fractures. The tiltmeter data recorded during the trials and later during preconditioning, indicated dips of 2° to 20°, which provided additional assurance that the fractures were essentially horizontal, especially at sites where no other monitoring was available. But attempts to analyse the tilt data for indications of asymmetric growth proved unreliable because the dip and dip direction are coupled to the location of the centre of fracture volume.

The theory of closely spaced fracture growth, developed using a 2D numerical model has been further verified by the measurments made during this project. The theory predicts that for the conditions at the Narrabri Coal site, hydraulic fractures placed sequentually at 1.25 m along a vertical borehole will grow with negligable curving to distance of 30 m or more, allowing the conglomerate roof rock to be preconditioned and weakened by placing fractures through its thickness. This was found to be the case, based on direct measurement of fracture arrival depths in offset boreholes.

The conglomerate caved soon after the start of Longwall 101, demonstrating the effectiveness of the intensive preconditioning carried out.

Hydraulic fracturing can be used for preconditioning of strong roof sequences. When conditions allow horizontal fractures to be placed from vertical boreholes, the preconditioning can be carried out from the surface.

Acknowledgements

The work described in this paper was undertaken as part of the overall windblast management project funded by Narrabri Coal Operations. The authors thank Narrabri Coal, CSIRO, and SCT Operations for granting permission to publish these results.

Author details

R. G. Jeffrey[1], Z. Chen[1], K. W. Mills[2] and S. Pegg[3]

1 CSIRO Petroleum and Geothermal, Australia

2 SCT Operations Pty Ltd, Australia

3 Narrabri Coal Operations Pty Ltd, Australia

References

[1] Van As, A, & Jeffrey, R. G. Caving induced by hydraulic fracturing at Northparkes Mines. In: J Girard, M. Liebman, C. Breeds, and T. Doe (Eds), The Fourth North American Rock Mechanics Symposium. 31 July- 3 August, (2000). Seattle, WA, USA. Rotterdam: A.A. Balkema.

[2] Chacon, E, Barrera, V, Jeffrey, R, & Van As, A. Hydraulic fracturing used to precondition ore and reduce fragment size for block caving. In: A. Karzulovic and M.A. Alfaro (Eds), MassMin August, (2004). Santiago, Chile. Instituto de Ingenieros de Chile., 2004, 22-25.

[3] Bunger, A, Jeffrey, R, Kear, J, & Zhang, X. Experimental investigation of the interaction among closely spaced hydraulic fractures. In 45th US Rock Mechanics / Geomechanics Symposium. June, (2011). San Francisco, CA, USA. ARMA., 26-29.

[4] Su DWHMcCaffrey JJ., Barletta L., Thomas EP., and Toothman RC. Hydraulic fracturing of sandstone and longwall roof control- implementation and evaluation. In S.S. Peng, C. Mark, and A.W. Khair (Eds), 20th International Conference on Ground Control in Mining, August, (2001). Morgantown, W.V., USA., 7-9.

[5]   He, H, Dou, L, Fan, J, Du, T, & Sun, X. Deep-hole directional fracturing of thick hard roof for rockburst prevention. Tunnelling and Underground Space Technology, doi: 10.1016/j.tust.(2012). , 32, 34-43.

[6]   Jeffrey, R. G, & Mills, K. W. Hydraulic fracturing applied to inducing longwall coal mine goaf falls. In: J Girard, M. Liebman, C. Breeds, and T. Doe (Eds), The Fourth North American Rock Mechanics Symposium. 31 July- 3 August, (2000). Seattle, WA, USA. Rotterdam: A.A. Balkema.

[7]   Medhurst, T. Narrabri Coal Pty. Ltd. Longwall support geotechnical assessment. Report by PDR Engineers, May (2009). pp.(8760)

[8]   Mills, K. W. In situ stress measurement using the ANZI stress cell. In: K. Sugawara and Y. Obara (Eds). The International Symposium on Rock Stress, October, (1997). Kumamoto, Japan. Rotterdam: A.A. Balkema., 7-10.

[9]   Mills, K. W. Interpretation of in situ stress measurements conducted at the start of longwall 1 at Whitehaven. Report by SCT Operations Pty. Ltd., March (2011). pp.

[10] Gray, I. Narrabri coal in-situ stress test (IST). Report by Sigra Pty. Ltd., 8 May (2006).

[11] Lecampion, B, & Peirce, A. Multipole moment decomposition for imaging hydraulic fractures from remote elastostatic data. Inverse Problems, (2007). doi:10.1088/

[12] Davis, P. V. Surface deformation associated with a dipping hydrofracture. Journal of Geophysical Research, (1983). B 88(7), 5826-5834.

[13] Lecampion, B, Jeffrey, R, & Detournay, E. Resolving the geometry of hydraulic fractures from tilt measurements, Pure and Applied Geophysics, (2005). doi:10.1007/ s00024-005-2786-4., 2005(162), 12-2433.

[14] Chen, Z. R, & Jeffrey, R. G. Tilt monitoring of hydraulic fracture preconditioning treatments. In the 43rd U.S. Rock Mech. Symposium and 4th U.S.-Canada Rock Mech. Symposium, Asheville, NC, 28 June- 1 July, (2009).

[15] Olson, J, Du, Y, & Du, J. Tiltmeter data inversion with continuous, non-uniform opening distributions: A new method for detecting hydraulic fracture geometry. International Journal of Rock Mechanics and Mining Sciences, 34(3-4), 236.ee10. doi: 10.1016/S1365-1609(97)00120-2,(1997). , 1-236.

[16] Mills, K. W, Jeffrey, R. G, & Zhang, X. Growth analysis and fracture mechanics based on measured stress change near a full-size hydraulic fracture. In the 6th NARMS Symposium, GulfRock (2004). Houston, June, 2004., 6-10.

[17] Jeffrey, R. G, Settari, A, Mills, K. W, Zhang, X, & Detournay, E. Hydraulic fracturing to induce caving: fracture model development and comparison to field data. In: D. Elsworth, J. P. Tinucci, and K. A. Heasley (Eds), DC Rocks: rock mechanics in the national interest: 38th U.S. Rock Mechanics Symposium, Jul 7-10, (2001). Washington, D.C. Lisse, Netherlands: Swets & Zeitlinger, B., 251-260

[18] Kear, J, White, J, Bunger, A. P, Jeffrey, R, & Hessami, M. Three dimensional forms of closely-spaced hydraulic fractures, In: A.P. Bunger, J.D. McLennan, and R.G. Jeffrey (Eds), The International Conference for Effective and Sustainable Hydraulic Fracturing, May, (2013). Brisbane, Australia. InTech: Rijeka, Croatia., 20-22.

[19] Bunger, A. P, Zhang, X, & Jeffrey, R. G. Parameters effecting the interaction among closely spaced hydraulic fractures. SPE Journal, (2012). , 17(1), 292-306.

Estimation of the Impact of Mining on Stresses by Actual Measurements in Pre and Post Mining Stages by Hydrofracture Method–A Case Study in a Copper Mine

Smarajit Sengupta, Dhubburi S. Subrahmanyam,
Rabindra Kumar Sinha and Govinda Shyam

Additional information is available at the end of the chapter

Abstract

To sustain and increase the productivity in a large underground copper mine in India the management of the mine decided to design and develop stopes below the mined out area. For the design of the stopes a detailed stress measurement programme was carried out by hydrofracture method at different depths from the developments available near the proposed stope. The result indicated a post mining induced high stress tensor with the direction of the maximum compression (maximum principal horizontal stress) rotated 70- 750 from the pre-mining stress tensor and oriented almost transverse to the ore body as against sub parallel to the orebody during pre- mining stage. A 3-D numerical modeling of the mine with pre mining stress tensor as input parameter substantiated the field result at the post mining stage. The generation of post - mining stress helped in understanding the impact of mining on the stress and was used for design and sequencing of the stoping operation for the safe and optimum extraction of the ore.

1. Introduction

Knowing the post mining stress condition is always of interest to the mine designer ahead of designing a mining method in the non-mined areas. This knowledge helps them in the design of stopes, mining sequence and rock reinforcement for the extraction of ores economically and safely. Previous work has examined the impact of mining on stresses as revealed by actual

Estimation of the Impact of Mining on Stresses by Actual Measurements in Pre and Post Mining Stages by
Hydrofracture Method–A Case Study in a Copper Mine

63

measurements at the site and included the use of 3D numerical methods to understand the
impact vis a vis mining to help in the designing of openings below mined out areas (Whyatt-
JK, Williams-TJ, Blake. W (1995).

In this study, in a deep underground copper mine, stress measurements using the hydrofracture method were carried out in two stages. At the pre-mining stage, when only few developments were available and at the post mining stage from the developments between the
mined out area and the non-mined out area.

Stress data generated from the stress measurements produced a value for the mining induced
stress gradient (post mining) which was found to be totally different from the stress gradients
of the area measured in the pre mining stage. The orientation of the Maximum Horizontal
principal Stress was found to be perturbed and lying perpendicular to the strike of the orebody
as against parallel orientation found during pre -mining stage. To understand the impact of
the mining on the stresses a 3-D numerical modeling study was carried out using a boundary
element method. The initial stress ratio from the pre mining stage measurement was used with
gravity loading to account for the surface topography, which is hilly. Three observation points
were monitored for stress change in mining, resulting from excavation effects and this data
was found to be in agreement the measured induced stresses. The study results helped in the
design of stopes, mining sequences and rock reinforcement.

2. Background

Hindustan Copper Limited (HCL), a public sector undertaking under the administrative
control of the Ministry of Mines, is engaged in mining, beneficiation, smelting, refining and
casting of refined copper metal. HCL maintains focused on its mission and vision which
include increasing the ore production by three times over a decade and implementing
continuous improvement in productivity. To continue to achieve these goals, it has geared up
to tap the resources from the un-mined areas by designing stopes below the mined areas.

The present study was undertaken in Kolihan Copper Mine, an important captive underground mine of Kolihan Copper Complex of HCL and this mine is situated near the village of
Khetri, in the District Jhunjhunu, Rajasthan. The mine plan to develop stope blocks at lower
levels below the mined out areas to sustain and increase the productivity.

For the design of stopes, in-situ stress is one of the most important factors which dictates
the size of the stopes and the size of the pillars and the sequence of extraction. The main
host rocks of Kolihan mines are garnetiferous chlorite quartz schist, quartzite and amphibolite quartzite. The strike length of the ore body is 600 m with a width varying from 30 m
to 100 m and the ore dips steeply to almost vertical. The main mining method adopted is
Large Diameter Blast Hole Stoping. The mine extends from 486 ML to 0 ML. (Hindustan
Copper Limited internal notes)

A detailed stress measurement programme was undertaken before the commencement of any
stoping activity (pre- mining stage) between 486 mL and 184 mL for the determination of stress

around the mine openings. Three locations with different depths (different rock covers) were selected inside the mine and stress measurements were conducted inside boreholes drilled from development tunnels (cross cuts), using the hydrofracture method.

Mining up to 306 ML is complete and presently mining is active at 246 ML and 184 ML. Mine development has to commence at lower level soon, below 184 ML.(Figure 2.) Thus it was felt to undertake a stress measurement programme again below the mined out area (post mining stage) to find the impact of mining activities on the stresses. Three levels with different rock covers were selected, similar to what was done in the pre-mining stage and stress measurements were conducted inside boreholes using the hydrofrac method.

3. Geology and tectonics

3.1. Geology

The rock formations of the area belong to the Alwar and Ajabgarh series of the Delhi system and are younger than the Aravalli system. Both rock formations are highly deformed and metamorphosed. Rocks occurring at Kolihan mines are Amphibolite quartzite/garnet chloride with principal economic mineral is chalcopyrite. Strike of the formation is N 30°E - S 30°W, dipping 50^0 - 85^0 westerly (Fig 1)

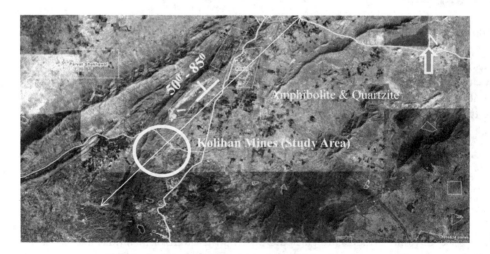

Figure 1. Geological and tectonic map of the project area

3.2. Tectonics

Structurally the thick prism of metasediments comprising rocks of Alwar and Ajabgarh series has been deformed into northeast –southwesterly trending longitudinal folds of large areal

Estimation of the Impact of Mining on Stresses by Actual Measurements in Pre and Post Mining Stages by
Hydrofracture Method–A Case Study in a Copper Mine

65

extent. In the northern part of the belt the simplest structures are represented by Khetri anticlines and synclines with increasing intensity of deformation. The simple structure passes westward into overturned Kolihan syncline which is slightly compressed in the north.

In the central part of the belt the formations show as anticline structures.

The southern part of the belt is separated from the central part by a major transverse fault. The southern part of the fault is marked by anticlines and synclines. The asymmetrically over-turned Kolihan syncline which is locally recumbent occupies a narrow zone. It plunged towards the SW and in the southern part the limbs are low dipping but gradually steepen northwards. The syncline is defined by the younger quartzites of the Ajabgarh series of reverse faulting (Dasgupta 1965).

4. Mining status

In the scheme of mining with respect to Kolihan Copper Mine the following methods have been adopted:

i.   Sub-level Open Stoping method

ii.   Blast Hole stoping Method

In the sub level open stoping method, sub levels are developed at vertical intervals of 18-20 m with a crown level at 9 m below uppermost levels. The size of the stope block is 30 m along strike which consists of 20 m of stope and 10 m of Rib Pillar.

In the blast hole stoping method a drill level is prepared below the crown pillar of 9 m. The size of the stope block is 30 m along the strike, which includes 16.6 m stope and 13.4 m Rib Pillar. The proposed stopes will be developed at the lower levels.

The mine extends from 486 ML to 0 ML with the surface RL of 486 m. Mining up to 306 ML is complete and presently it is active at 246 ML and 184 ML. Mine development has to commence at lower level soon.

5. Methodology

In-situ stress measurement using the hydrofracture method was carried out both during pre mining and post mining stages. Three boreholes were drilled, one each from 184 ML, 124 ML and 64 ML, for post mining stress determination.

The in-situ stress measurement was carried out by using HTPF (Hydraulic Tests on Pre existing Fracture) as introduced by Cornet et al. 1986]. The advantages of HTPF method are

i.   The boreholes are not required to be oriented along one of the principal stress direction like in classical methods

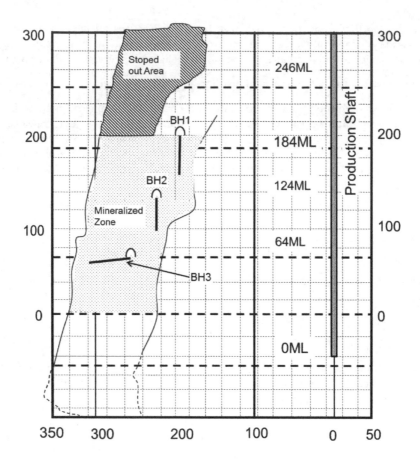

Figure 2. Status of Mining activities in Kolihan mine (ML= Meter level which indicates altitude from mean sea level)

ii.   A new induced fracture is not essentially required to be created for stress evaluation. Stress can be evaluated both from preexisting/induced fractures

A schematic diagram showing set up of the hydrofracture system assembly is shown in Fig.3.

The straddle packer assembly (Hydrofrac assembly Fig 4) was used for fracture initiation/ opening and further extension. The straddle packer assembly consisted of a test interval of length 200 mm and two 250 mm steel reinforced packer (42 mm dia, burst pressure = 70 MPa) units attached at either end of the test interval. In the case of hydrofrac experiments in the 48 mm diameter boreholes at the present Project, the straddle packer unit was operated by 1500 mm long and 32 mm diameter tubes (dual line packer inflation + injection unit combined in one). The maximum injection rate of the electrically driven pump was 10 lit /min using water for pressurisation. All the events of injection were recorded in continuous real time digital mode.

Estimation of the Impact of Mining on Stresses by Actual Measurements in Pre and Post Mining Stages by
Hydrofracture Method–A Case Study in a Copper Mine

67

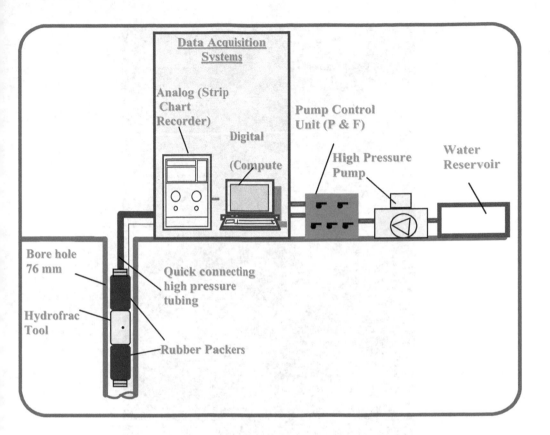

Figure 3. Schematic diagram of Hydrofrac Experiment Set-up

After all the hydraulic fracturing tests were conducted in all the boreholes, an impression packer tool with a soft rubber skin together with a magnetic single shot orientation device was run into the holes to obtain information on the orientation of the induced or opened fracture traces at the borehole wall.

Two data analyses programmes were used in the analyses. They are called Plane and Gensim.

The *software Plane* incorporates the impression data with the compass data as input parameters and gives the strike, dip and dip direction (fracture orientation data) as the output.

The Software Gensim computes the stress field on the basis of measured shut in pressure and fracture orientation data. The vertical stress is assumed to be a principal stress and its magnitude is taken as equal to the weight of the overburden. The powerful Gensim programme requires only the shut in pressure and the orientation of an induced or pre-existing fracture

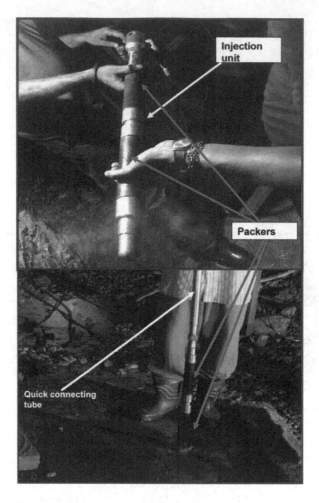

Figure 4. Hydrofracture equipment used

6. Stress evaluation procedures and results

The in-situ stress measurement were made from inside two vertical and one horizontal boreholes drilled from three levels. Tests were conducted with the following situations:

i.   Presence of anisotropic rock.

ii.   Presence of mining induced stress.

Due to the above aspects a medium to large scatter in fracture orientation data were noticed which negated the use of classical simple hydrofrac hypothesis suggested by Hubert and Wills (1957). Therefore data analysis required a more sophisticated meth-

Estimation of the Impact of Mining on Stresses by Actual Measurements in Pre and Post Mining Stages by
Hydrofracture Method–A Case Study in a Copper Mine

69

od, namely the interpretation of measured normal stress acting across arbitrary oriented fracture planes.

In this method the shut-in pressure P_{si} is used to measure the normal stress component under the assumption that the vertical stress is a principal stress axis and the vertical stress magnitude σ_V is equal to the weight of the overburden.

The analysis program *GENSIM* was used to calculate the magnitude and the direction of principal stresses on the basis of the following equation:

$$\sigma_h = (P_{si} - n^2.\sigma_V) / (m^2 + 1^2.\sigma_H \mathit{b}_h) \tag{1}$$

Where, l, m, n is the cosines of the direction of the induced fracture plane related to the principal stress axis.

The calculations involve obtaining the best fit based on using all shut-in pressure data derived from the measurements in the boreholes and varying the ratio σ_H/σ_h and the strike direction of σ_H.

The pre-mining and post mining stress tensors as revealed are given in tables 1 and 2

Principal Stresses			
σ_V MPa	σ_H MPa	σ_h MPa	Rock Cover Depth m
6.97	8.4	5.6	203
7.88	8.89	5.93	268
10.7	12.65	7.7	364

Table 1. Pre mining stress tensor as revealed by hydrofrac stress

Principal stresses	184 ML	124 ML	64 ML
Rock cover	195m	184 m	530 m
Vertical Stress (σ_V) MPa (2.7 gm/cc + 1.4 gm/cc density of solid and loose rocks respectively)	9.28	9.89	14.02
Maximum Horizontal principal Stress (σ_H) in MPa	21.78	22.78	23.94
Minimum Horizontal principal Stress (σ_h) in MPa	10.89	11.39	15.96
Maximum Horizontal principal Stress direction	N 80°	N 80°	N 90°
$K = \sigma_H/\sigma_V$	2.35	2.30	1.71

Table 2. Post mining stress tensor as revealed by hydrofrac stress

Table 3 shows the comparison of pre and post mining stress gradient

Stresses	Pre – Mining Stage (486 mL to 184mL)	Post Mining Stage (184 mL to 0 mL)	Remarks
Maximum Horizontal principal Stress (σ_H) orientation	N 10° to N 20°	N 85° to N 90°	Rotation of horizontal stress orientation due to stoping
Stress gradient (σ_H)	0.031 Z +1.5968 $R^2 = 0.91$	0.0048 Z + 21.379 $R^2 = 0.7627$	Change in stress gradient due to mining
Stress gradient (σ_h)	0.0145 Z +2.3892 $R^2 = 0.93$	0.01437 Z + 8.412 $R^2 = 0.9862$	Change in stress gradient due to mining

Table 3. Comparison between pre and post mining stress gradient

7. Numerical modeling

A numerical modeling was carried out using the boundary element method to understand post mining induced stresses vis a vis mining. The initial stresses gradient of the pre mining stage was used with gravity loading as the surface topography is hilly. Three observation points were monitored for stress change in mining, due to excavation effects. The stress contour of the model is shown in figure 5

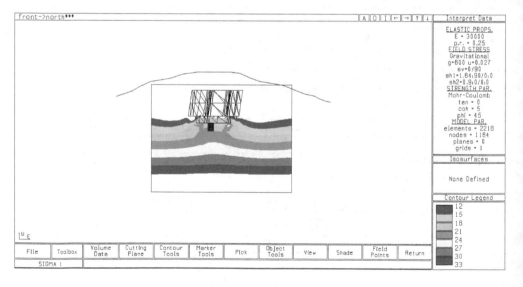

Figure 5. Major principal stress contour of the modeled stope.

The results of the stress output as revealed by the numerical model are given in Table 4.

ML	Sigma 1			Sigma 2			Sigma 3		
	Magnitude	Dip	Direction	Magnitude	Dip	Direction	Magnitude	Dip	Direction
ML - 184	21.15	7.32	272.59	10.9	30	178	6.99	58.25	14.59
ML - 124	23.33	2.32	92.52	11.48	10.29	182.94	6.36	79.43	349.94
ML- 64	24.66	6.3	90.43	13.07	12.28	181.81	10.47	76.14	333.81

Table 4. Stress magnitude and orientation as revealed by numerical model

The modeling studies reveal that the measured value of the stresses agree reasonably with the computation values which is compared in Table 5

Stresses	Post Mining Stage (184 mL to 0 mL)	Numerical modelling
Maximum Horizontal principal Stress (σ_H) orientation	N 85^0 to N 90^0	N 90^0 to N 92^0
Stress gradient (σ_H)	0.0048 Z + 21.379 $R^2 = 0.7627$	0.0069 Z + 20.924 $R^2 = 0.5943$
Stress gradient (σ_h)	0.01437 Z + 8.412 $R^2 = 0.9862$	0.0055 Z + 10.158 $R^2 = 0.9188$

Table 5. Stress magnitude and orientation as revealed by numerical model

8. Discussion and conclusion

The availability of stress results during pre - mining stage and subsequent measurement of stresses at the post mining stage has refined our understanding of the in-situ stress vis a vis mining. The change in the orientation of the major compression from a favourable N10-20 ^{0} (Strike of ore body N 30^0 and crown pillar oriented parallel to ore body) during pre- mining stage to unfavourable N85-90 ^{0} at the post mining stage has prompted to redesign the stopes and support systems below the mined out area.

Acknowledgements

We are thankful to the Director National Institute of Rock Mechanics, India for the permission to publish the work. The authorities and staffs of Hindustan Copper limited are also thankfully acknowledged.

Author details

Smarajit Sengupta, Dhubburi S. Subrahmanyam, Rabindra Kumar Sinha and
Govinda Shyam

National Institute of Rock Mechanics, Bengaluru, India

References

[1] Whyatt-JKWilliams-TJ, Blake. W ((1995). In-Situ Stress in Lucky Friday mines W Reference:U.S. Department of the Interior, Bureau of Mines, Report of Investigations 9582. NTIS stock (PB96-131685)

[2] Hindustan Copper Limited Internal Notes

[3] Dasguupta, B. (1965). Khetri Copper Belt. GSI memoirs, , 98

[4] Cornet, F. H. (1986). Stress determination from Hydraulic Tests on Pre-exiting Fractures- the HTPF Method. Proc. Intl Symp, Rock Stress and Rock Stress Measurements, CENTEK Publ., Lulea, , 301-311.

[5] Hubbert, K. M, & Willis, D. G. (1957). Mechanics of Hydraulic Fracturing, Petroleum Transactions AIME, T., 210, 4597.

Thermo-Hydro-Mechanical Systems

Chapter 5

Thermal Effects on Shear Fracturing and Injectivity During CO$_2$ Storage

Somayeh Goodarzi, Antonin Settari,
Mark Zoback and David W. Keith

Additional information is available at the end of the chapter

Abstract

With almost two hundred coal burning power plants in Ohio River valley, this region is considered important for evaluation of CO$_2$ storage potential. In a CO$_2$ storage project, the temperature of the injected CO$_2$ is usually considerably lower than the formation temperature. The heat transfer between the injected fluid and rock has to be investigated in order to test the viability of the target formation to act as an effective storage unit and to optimize the storage process. In our previous work we have introduced the controversial idea of injecting CO$_2$ for storage at fracturing conditions in order to improve injectivity and economics. Here we examine the thermal aspects of such process in a setting typical for Ohio River Valley target formation.

A coupled flow, geomechanical and heat transfer model for the potential injection zone and surrounding formations has been developed. All the modeling focuses on a single well performance and considers induced fracturing for both isothermal and thermal injection conditions. The induced thermal effects of CO$_2$ injection on stresses, and fracture pressure, and propagation are investigated. Possibility of shear failure in the caprock resulting from heat transfer between reservoir and the overburden layers is also examined.

In the thermal case, the total minimum stress at the wellbore decreases with time and falls below the injection pressure quite early during injection. Therefore, fracturing occurs at considerably lower pressure, when thermal effects are present. The coupled thermal and dynamic fracture model shows that these effects could increase the speed of fracture propagation in the storage layer depending on the injection rate. These phenomena are dependent primarily on the difference between the injection and reservoir temperature.

Our results show that shortly after injection, the induced expansion in caprock lead to slight increase of total stresses (poroelasticity) which will reduce the chance of shear failure. However as soon as total minimum stress in the caprock decreases due to thermal diffusion between the reservoir and caprock, thermoelasticity dominates and the chance of shear failure increases in the caprock.

Incorporation of thermal effects in modeling of CO_2 injection is significant for understanding the dynamics of induced fracturing in storage operations. Our work shows that the injection capacity with cold CO_2 injection could be significantly lower than expected, and it may be impractical to avoid induced fracture development. In risk assessment studies inclusion of the thermal effects will help prevent the unexpected leakage in storage projects.

1. Introduction

Past storage pilot projects and enhanced oil recovery efforts have shown that, geologic sequestration of CO_2 is a technically viable means of reducing anthropogenic emission of CO_2 from accumulating in the atmosphere [1,2,3]. Security of storage is one of the most important concerns with the long term injection of CO_2 in underground formations. Injection of CO_2 induces stress and pore pressure changes which could eventually lead to the formation or reactivation of fracture networks and/or shear failure which could potentially provide pathways for CO_2 leakage through previously impermeable rocks [4]. Therefore geomechanical modeling plays a very important role in risk assessment of geological storage of CO_2.

In order to determine whether the induced stress changes compromises the ability of the formation to act as an effective storage unit, a geomechanical assessment of the formation integrity must be carried out. In our previous work, we have studied the dynamic propagation of fracture in the Rose Run sandstone reservoir in Ohio River valley under isothermal [5] and thermal condition [6] for injection above fracture pressure. In this paper, the thermal effect of injection on the possibility of tensile and shear failure in the reservoir and caprock are studied for injection below fracture pressure. This study utilized a fully coupled reservoir flow and geomechanical model which allows accounting for poroelastic and thermoelastic effects and can model static and/or dynamic fractures.

To examine the possibility of shear failure in the caprock, Mohr-Coloumb Criteria was used.

2. Construction of the flow, thermal and geomechanical model

A coupled flow, thermal, and geomechanical model has been developed in order to study the thermo-elastic and poro-elastic response of the injection and surrounding layer to increasing of pressure and reduction of temperature after CO_2 injection. Ohio River valley is located in a relatively stable, intraplate tectonic setting and the regional stress state is in strike slip faulting regime with the maximum stress oriented northeast to east-northeast [7].

This study used the fluid and rock mechanical properties provided by Lucier et al. [8]. The stratigraphic sequence of the geological layers in the study area and the relative location of the potential injection layer, Rose Run Sandstone (RRS) at the Mountaineer site is shown in Figure 1. RRS has an average thickness of 30 m and is extended from 2355-2385 m. The direction of maximum and minimum horizontal stresses is reported to be in N47E(±13) and N43W (±13) respectively [8]. All the models in this study are aligned along these directions, in order to avoid having initial non-zero value of shear stresses in principal stress directions.

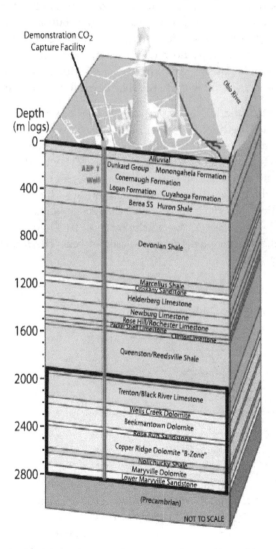

Figure 1. Generalized stratigraphy of the study area at the Mountaineer site. The well location and the general stratig-raphy intersected by well is illustrated in the picture. The black box shows the boundaries of the area of previous work by Lucier et al., [8], Modified from [9]

The developed element of symmetry model that covers 8000x8000x2575 m of study area, has 50x50x9 grid block in x, y and z directions respectively. The injection well is located at the top left corner of this model. RRS was gridded into three layers with 5, 10 and 15 m thickness. The adjacent Beekmanton Caprock was refined into 3 layers (10, 50 and 126 m) to capture and predict the potential growth of fracture through this layer (and the resulting possibility for CO_2 leakage). The horizontal and vertical permeability of the caprock layers in the model are given as 2E-10 and 1E-10 md resepectively. Average properties of 5%, 20 md and 10 md for porosity, horizontal and vertical permeability were given to the injection layer. These values are the probability averages of the given property distributions for Rose Run sandstone formation [8]. The initial pressure and temperature of the RRS is 26000 kPa and 63.1 C. The fluid flow is modeled by two-phase flow with dissolution of CO_2 in water. Van Genutchen function with an irreducible gas saturation of 0.05, an irreducible liquid saturation of 0.2 and an exponent of 0.457 was used to generate relative permeability data [10].

The mechanical properties and initial stress profile is required to be added to the geomechanical model and coupled with the flow model in order to be able to study the mutual effect of pressure and stresses and the resulting effect on fracture propagation and injectivity. The mechanical properties for this model are listed in Table 1. The listed value with the exception of grain Modulus are all extracted directly from Lucier et al. paper [8]. Grain modulus was back- calculated from the given Biot constants and Young's Modulus. The Biot constant α is important for computing the effects of pressure changes on stress. At the Mountaineer site, Lucier et al. estimated α to be very low - in the range of 0.03 to 0.2. In this analysis, a mean value of 0.11 was used to calculate the poroelastic effects. The formation rock density is assumed to be 2500 kg/m^3 [8].

Layer-top depth (m)	Thickness (m)	Young's Modulus (kpa)	Poisson's Ratio	Grain Modulus (Kpa)
Shale-Surface	1911	6.00E+07	0.29	5.25E+07
Limestone-1911	253	7.05E+07	0.3	6.61E+07
Dolomite-2164	186	8.96E+07	0.28	7.51E+07
Rose Run Sandstone-2350	30	8.73E+07	0.25	6.53E+07
Dolomite-2380	195	9.47E+07	0.28	8.05E+07

Table 1. Rock Mechanical properties of the coupled model

The initial pressure, horizontal and vertical stress profile for different depths in Ohio River Valley is shown in Figure 2. It is important to note that the horizontal stresses are lower in RRS (the injection layer) than in the surrounding layers. This is a common behavior due to generally having larger Poisson's ratio for the surrounding layers than the reservoir. In many situations the stresses in caprock (low permeability rock) are larger than in the reservoir (permeable formations), because of differences in Poisson's ratio, material properties, stress history and other factors. This is well documented in hydraulic fracturing literature and is the primary mechanism for containment of fractures to the target zone. This initial stress contrast is very critical when considering fracture propagation in the reservoir layer for enhancing injectivity while avoiding the risk of fracture growth through upper caprock layers. As mentioned before, since the temperature of injected CO$_2$ (at approximately 30 deg C) is smaller than the formation temperature (at 60 deg C), thermal effects of injection on fluid flow and geomechanics must be included in the model. This coupling is achieved by solving the energy balance equation within the fluid flow model, and including the thermoelasticity term in the geomechanical model (included in the constitutive model of the rock).

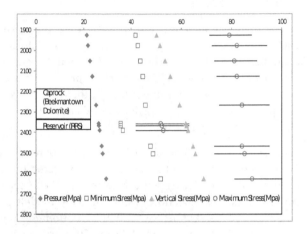

Figure 2. Initial pressure, horizontal and vertical stress profile in Ohio River Valley [8]

The average thermal properties for the rock, as well as injected and in-place fluids used for this study are listed in Table 2 [8,11-14].

	Rock	**Water**	**CO$_2$**
Volumetric Thermal Expansion Coefficient (1/deg K)	5.4E-6	2.1E-4	3.003E-3
Heat Capacity(Kj/Kg deg K)	0.9	4.182	0.84
Thermal Conductivity(W/ m deg K)	2.34	0.65	0.084

Table 2. Thermal properties of fluids and rock

The boundary condition for the fluid flow model is that there is no flow across the boundary of the model. The constraints for the geomechanical model are as follows. The right and left sides of the model are fixed in the x-direction so there would be no displacement in the x-direction. The front and back sides of the model are fixed in Y direction. The bottom side of the model is fixed in vertical direction and the top of the model is free to move in all directions. Stresses were initialized according to data in Fig. 2. All injections are done through a single vertical well with constant injection rate.

3. Thermal fracturing in the reservoir below isothermal fracture pressure

Thermal effects of CO_2 injection is expected to affect the magnitude of displacements, pressure, stresses, and the possibility of shear and tensile failure in the reservoir and caprock. Injecting fluid with temperature lower than reservoir rock temperature will cause reduction of stresses in the injection layer and once the temperature front has reached a relatively large area around the wellbore, this reduction in stresses will result in negative volumetric strains that can propagate to the surface. Therefore the surface displacement for the thermal model would be smaller than that of isothermal model [6].

One of the most important effects of injecting a fluid with a lower than reservoir temperature is the reduction of fracture pressure. Cooling of the formation rock during injection of cold CO_2 through thermal conduction and convection lowers the total stresses in the reservoir and possibly caprock layer. This results in reduction of fracture pressure and the pressure differential available for injection, and therefore injectivity. In the case of injection at fracturing conditions, the fracture propagation pressure will decrease and, if the same injection rate is used, this will accelerate fracture propagation.

In order to examine thermal effects of injection on the possibility of reaching tensile failure in the reservoir, the variation of total stress and pressure needs to be studied. In order to do that, the coupled geomechanical, flow and thermal simulation has been carried out with two different injection temperatures. The injection of CO_2 for these models is through a single vertical well with constant rate of 3.4E4 m3/day such that the bottomhole pressure will remain below fracture pressure for the isothermal model during 30 years of injection. It should be noted that fracturing was not allowed in these models. Thermal model in this study refers to injection temperature (30 C) being lower than the reservoir temperature (60 C),while in the isothermal model, it is equal to reservoir's temperature. Figure 3 shows the modeling results for pressure, and total minimum stress for well block in the reservoir during 30 years of injection for the thermal and isothermal model. As it is seen in Figure 3, the total stress falls below the bottomhole pressure (fracture pressure) for the thermal model in the reservoir at quite early injection times which means that minimum effective stress will reduce beyond zero. Therefore, although CO_2 is injected below the original fracture pressure, fracture would initiate in the reservoir for the thermal case. Since fracturing is not allowed in these models, the stress magnitudes after the onset of fracturing are not valid. If propagation was allowed, minimum total stress would reduce to slightly below bottomhole pressure such that the effective stress on the fracture wall would remain close to zero.

In order to study the thermal effects of injection on the propagation of the induced fracture, for the next set of model cases we allowed fracture propagation in all layers. To model the potential fracture propagation, a transmissibility multiplier technique is incorporated in the model, which essentially accounts for the fluid flow transmissibility through the fracture by a transmissibility multiplier function, specified as a table. The multiplier is calculated from an estimated fracture opening of a 2-D Griffith crack [15] based on the mechanical properties of the injection zone and an estimate of the fracture height [5]. This function can be incorporated

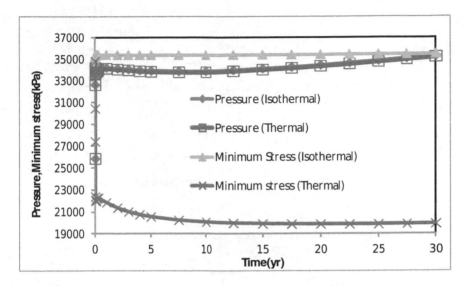

Figure 3. Pressure, minimum stress, and stress level for the thermal and isothermal model for the well block in the reservoir's top layer

in the model both as a function of pressure or effective minimum stress. The actual fracture geometry can be calculated by the coupled model.

Figure 4 shows the bottomhole pressure and fracture length for the reservoir's top layer for the thermal and isothermal model. As expected, since the bottomhole pressure remains below the fracture pressure for the isothermal model, there is no fracture initiation in the reservoir for the isothermal model. However since minimum effective stress reduces beyond zero in the thermal model (thermoelastic effects), fracture initiates and propagates through reservoir to a half length of 250 m. The bottomhole pressure in the thermal model is now significantly different. For the thermal model, it increases to fracture initiation pressure (equal to the thermally reduced minimum total stress) and then remains almost constant for the injection period. However for the isothermal model, the pressure history is the same as in Fig. 4.

Figure 5 shows the fracture length, pressure and temperature profile for the well block in the reservoir's top and bottom layer. The results show that under thermal conditions, fracture propagates to a larger extent in the lower reservoir layers than the top ones. As seen in the Figure, the pressures in the reservoir's top and bottom layers are very close. However, the temperature in the bottom of the reservoir is significantly lower than in the top. This results in higher reduction of minimum total stress and lower fracture pressure for the reservoir's bottom layer. This effect can also be clearly seen in Figure 6 which shows the permeability multiplier, temperature and pressure profile in fracture plane near the wellbore (zoomed to 300 m) across reservoir layers. As seen the temperature reduction and permeability multipliers are higher in the bottom layer.

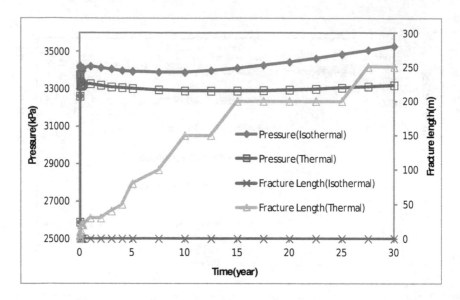

Figure 4. Bottomhole pressure and fracture length for the reservoir's top layer for the thermal and isothermal model

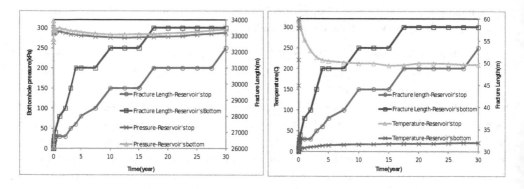

Figure 5. Fracture length and pressure profile (left), fracture length and temperature profile (right) for the reservoir's top and bottom layer

4. Thermally induced shear failure in the caprock

The thermally induced reduction of minimum stress in the caprock could lead to tensile or shear failure of the formation rock which could cause tensile fracture propagation through these layers or hydraulic communication through shear fractures. If there is no stress contrast between the caprock and reservoir, reaching tensile failure in the caprock is possible for injection above fracture pressure [16]. In this study, since the horizontal stresses in the caprock

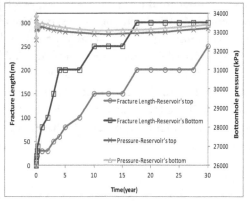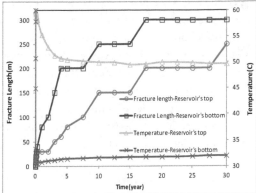

Figure 6. Permeability multiplier(top), Temperature(middle), Pressure distribution (bottom) after 30 years of injection for the reservoir layers

are significantly larger than in the reservoir, tensile fracturing in the caprock is of low likelihood. However given the large initial deviatoric stress in the caprock, the chance of reaching shear failure due to thermally induced stresses is high. In order to evaluate the possibility of reaching shear failure, we have used the Mohr-Coloumb criteria and studied the variation of "Stress level", l_σ in the caprock during injection. Stress level is defined as the ratio of deviatoric stress at the current condition to the deviatoric stress at failure condition:

$$l_\sigma = \frac{\sigma'_{dev}}{(\sigma'_{dev})_f} = \frac{(\sigma'_{max} - \sigma'_{min})}{(\sigma'_{max} - \sigma'_{min})_f} \leq 1 \tag{1}$$

Where, l_σ is the stress level, σ_{dev}' is the deviatoric stress at the current condition, $(\sigma_{dev}')_f$ is the deviatoric stress at failure, σ_{max}' is the maximum principal stress, σ_{min}' is the minimum principal stress (all stresses are effective). The deviatoric stress at failure is a function of cohesion c and friction angle ϕ and is defined as:

$$(\sigma'_{dev})_f = \frac{2cCos\varphi + 2\sigma'_3 Sin\varphi}{(1 - Sin\varphi)} \tag{2}$$

When the stress level is less than 1, the shear stress has not exceeded the shear strength of the rock and when it is larger than 1, the shear strength of the rock has been reached in a plane which is aligned in the direction found from the Mohr stress circle. The nominal rock cohesion for the caprock (Beekmantown Dolomite) is 9000 kPa [17]. Linear elastic constitutive model was used to describe the mechanical behavior of the formation rock. In order to examine the thermal effects on the stress state in the caprock, the variation of total stress, pressure and stress level needs to be studied.

Figure 7 shows the pressure, stress and stress level evolution for the well block in the caprock for the thermal and isothermal model. In the isothermal model, due to the low permeability of caprock, pressure increase in caprock is negligible compared to the reservoir, and stress level remains low. However as seen in Figure 8 (which shows the stress, pressure and temperature variation of the well block in the caprock during the first 10 years of injection), the first caprock layer is quickly pressurized, and later its temperature also decreases due to heat transfer. Stress level is rapidly increasing with time due to thermally induced decrease of total stresses. Therefore the chance of failing the rock in shear for the caprock is higher for the thermal model compared to isothermal model.

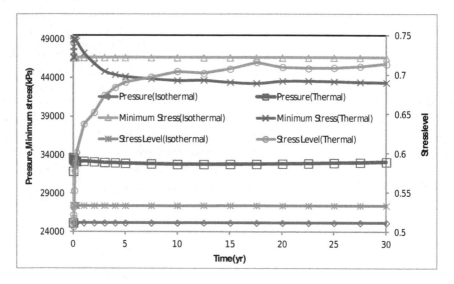

Figure 7. pressure, minimum total stress, and stress level for the thermal and isothermal model for the well block in the immediate caprock layer

The changes in the stress level correspond to the movement of the Mohr circle with time. Shortly after injection (0.1 days), the stress circle moves to the right due to the slight growth of total stresses. This is a poroelastic effect which is a result of early time-increase of the block pressure in the caprock. This can be clearly seen in Figure 8. However, as soon as the block temperature is lowered due to thermal diffusion (conduction), thermoelasticity dominates and total minimum stress reduces (Figure 8) and stress circle moves to the left toward the failure cone.

The mechanism shown here is somewhat exaggerated because of the upstream numerical treatment of the fracture transmissibility between the blocks, but the relative comparison is valid. Accurate modeling would require very fine vertical grid at the reservoir-caprock interface or the development of more sophisticated numerical technique. These aspects are being currently studied.

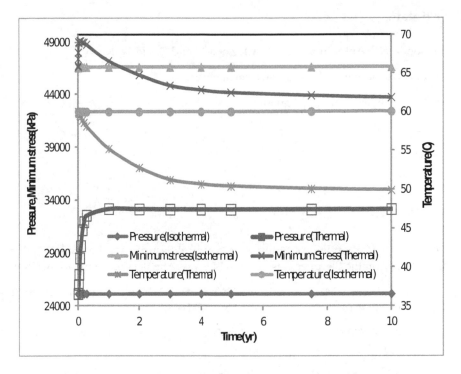

Figure 8. Minimum total stress, pressure and temperature variation for the well block in the caprock during the first 10 years of CO$_2$ injection

5. Conclusions

This paper studies thermo-elastic and poro-elastic response of the reservoir and caprock to increasing of pressure and reduction of temperature after CO$_2$ injection and the resulting consequences for the possibility of reaching tensile or shear failure both for the injection below and above reservoir's fracture pressure.

When injecting a fluid below isothermal fracture pressure with a temperature below reservoir temperature, the fracture pressure will decrease and minimum effective stress in the reservoir may reduce below zero for the fracturing to initiate and propagate in the reservoir.

Our results show that the reduction of the minimum effective stress due to thermal effects is larger for the lower reservoir layers. Therefore in case of dynamic fracture propagation, fracture growth would be larger for the lower reservoir layers due to larger cooling for these layers.

Thermal effects of injection with cold CO$_2$ may also create the possibility of shear failure in the caprock.

Author details

Somayeh Goodarzi[1*], Antonin Settari[1], Mark Zoback[2] and David W. Keith[3]

*Address all correspondence to: sgoodarz@ucalgary.ca

1 University of Calgary, Calgary, Canada

2 Stanford University, USA

3 Harvard University, USA

References

[1] Solomon, S. Carbon Dioxide Storage: Geological Security and Environmental Issues-Case Study on the Sleipner Gas Field in Norway. The Bellona Foundation. (2006).

[2] Preston, C, Monea, M, et al. IEA GHG Weyburn CO_2 Monitoring and Storage Project, Fuel Processing Technology. (2005). , 86, 1547-1568.

[3] Wright, L. W. (2007). The In Salah Gas CO2 Storage Project. International Petroleum Technology Conference

[4] QuintessaNational Institute of Advanced Industrial Science and Technology of Japan, Quintessa Japan, JGC Corporation, Mizuho Information and Research Institute. (2007). Building Confidence in Geological Storage of Carbon Dioxide. Ministry of Economy, Trade and Industry (METI), IEA Greenhouse Gas R&D Programme (IEA GHG).

[5] Goodarzi, S, Settari, A, Zoback, M, & Keith, D. A Coupled Geomechanical Reservoir Simulation analysis of CO_2 storage in a Saline Aquifer in the Ohio River Valley. (2011). Environmental Geosciences journal. American Association of Petroleum Geologists. , 18(3), 1-20.

[6] Goodarzi, S, Settari, A, Zoback, M, & Keith, D. Thermal Aspects of Geomechanics and Induced Fracturing in CO_2 Injection With Application to CO_2 Sequestration in Ohio River Valley. SPE-PP, SPE International Conference on CO_2 Capture, Storage, and Utilization held in New Orleans, Louisiana, USA, 10-12 November (2010). , 139706.

[7] Zoback, M. D, & Zoback, M. L. State of stress and intraplate earthquakes in the central and eastern United States. Science, (1981). , 213, 96-109.

[8] Lucier, A, Zoback, M, Gupta, N, & Ramakrishnan, T. S. (2006). Geomechanical aspects of CO_2 sequestration in a deep saline reservoir in the Ohio River Valley region. Environmental Geosciences 13 (2), 85-103.

[9] Gupta, N. (2008). The Ohio river valley CO$_2$ storage project, Final Technical Report, prepared for US Department of Energy-National Energy Technology Laboratory

[10] Van Genuchten, M. T. equation for predicting the hydraulic conductivity of unsaturated soils: Soil Science Society of America Journal, , 44, 892-898.

[11] Collieu, A. McB., Powney, D. J., Girifalco, L. A. et al., (1975). The Mechanical and Thermal Properties of Materials and Statistical Physics of Materials. Phys. Today 28, 51

[12] Fjaer, E, Holt, R. M, Horsrud, P, et al. (2008). Petroleum Related Rock Mechanics. 441. Elsevier.

[13] Guildner, L. A. (1958). The thermal conductivity of carbon dioxide in the region of the critical point, Proceedings of the National Academy of Sciences of the United States of America, , 44(11), 1149-1153.

[14] Yaws, C. (2008). Thermophysical properties of chemicals and hydrocarbons. 793. William Andrew Publishing

[15] Sneddon, I. N, & Lowengrub, M. Crack Problems in the Classical Theory of Elasticity, John Wiley & Sons Inc., New York, (1969). , 19.

[16] Goodarzi, S, Settari, A, & Keith, D. (2012). Geomechanical modeling for CO2 storage in Nisku aquifer in Wabamun lake area in Canada. International Journal of Greenhouse Gas Control , 10, 113-122.

[17] Maurer, W. C. (1965). Bit-Tooth Penetration under Simulated Borehole Conditions, Petroleum Transactions

An Efficient and Accurate Approach for Studying the Heat Extraction from Multiple Recharge and Discharge Wells

Bisheng Wu, Xi Zhang, Andrew Bunger and
Rob Jeffrey

Additional information is available at the end of the chapter

Abstract

In order to understand the thermal recovery Behavior of an engineered geothermal system (EGS), this paper develops a model in which fluid circulates in a single, planar hydraulic fracture with a constant hydraulic aperture via multiple recharging and discharging wells. The coupled equations for heat convection in the fracture plane and heat transfer into the rock are provided for steady and irrotational fluid flow conditions. By using velocity potentials and streamline functions, the temperature along a streamline is found to be only a function of the potential. By utilizing the Laplace transformation, the analytical solutions in the Laplace space for the temperature field are found, which are numerically inverted for time-domain results. Several examples with different arrangements of injection and production wells are investigated and the comparison with other published results is provided. The semi-analytical results demonstrate that the proposed model provides an efficient and accurate approach for predicting the temperatures of a multi-well reservoir system.

1. Introduction

Heat contained in the upper 10 km of the Earth's crust represents a large, accessible, low-emission energy source that can substitute for other energy sources that produce significantly more greenhouse gas. Geothermal energy has become one of the most promising energy alternatives in the future. For example, Australia has a large volume of identified high heat producing granites within 3 to 5km of the surface. In some places at 5kms the temperature is more than 250^0 C. One cubic kilometer of hot granite at 250^0 C has the stored energy equivalent of 40 million barrels of oil (Geodynamics website). Therefore, developing methods to capture this resource of clean energy will help realize the potential of commercial EGS. In addition to the physical experiments and field exploitation, theoretical studies such as mathematical modeling are important in developing ways to maximize heat extraction from geothermal reservoirs treated by hydraulic fracturing.

Mathematical modeling of thermal problems associated with oil recovery or geothermal heat extraction from reservoirs containing hydraulic or natural fractures has been studied extensively. The models can be classified into three types based on the dimension of the heat transfer problem. The first type is based on one-dimensional (1-D) heat diffusion in the fracture and it mainly contains two cases. The first one considers 1D fluid flow in a 2D fracture. In this case, the fluid velocity is uniform and no singular point in pressure exists at the injection or production well [1-3]. The second one considers fluid flow that is radial and axisymmetric, for example the fluid is injected into a single well and produced from a ring of production wells that are all at the same radial distance from the injection well [4]. Due to the symmetry, the problem can be treated as one-dimensional based on the radial distance. The fluid velocity can be obtained either by considering the effect of the wellbore size or neglecting it. When the injection or pumping at the well is regarded as a point source or sink, the velocity will have a singularity of 1/r theoretically at the injection and production points. Fortunately, this singularity is overcome by the geometrical symmetry [5, 6] for the case with a single well.

The second type is so-called $2^1/_2$ dimensional model [7] which results from 2-D heat transfer within fractures and 1-D heat transfer within the adjacent rocks, it becomes more difficult to find analytical solutions for these cases because of the coupling of the steady fluid flow and heat transfer. There are mainly two ways to obtain the fluid velocity analytically for the cases with single, dipole or multiple wells; one method calculates the pressure by using a Green's function [8], and the other method calculates the velocity potential and stream functions by using the source solution for the 2-D Laplace equation [9, 10]. From the perspective of modeling the long lifetime of a geothermal reservoir, the above two approaches are functionally the same. Although the first method can obtain accurate results for the fluid pressure and thus the fluid velocity, there still exists two coordinates in the governing equation and the difficulty of solving this equation is not reduced. In addition, the singularity issues make it invalid to use the boundary condition (i.e. injection temperature) when solving the equations. For this case, solutions still require the use of numerical methods.

In order to overcome the above issues including the computational difficulty and singularity at the source or sink points, Muskat [11] proposed a method in which an orthogonal set of curvilinear coordinates is introduced that correspond to the permanent set of equipotential and streamline surfaces. By using this transformation, the terms involving cross products related to the fluid diffusion and heat transfer are greatly simplified. Gringarten and Sauty [12] were the first to apply this concept to solve the dipole well problem of fluid flow and heat diffusion in an infinite fracture by using velocity potentials and stream functions based on the results of Dacosta and Bennett [9]. In their model, each channel leaving a particular injection well is treated separately and the two-dimensional heat diffusion problem in the fracture plane is simplified to be one-dimensional. Later Rodemann [13], Schulz [14], Heuer et al. [15] and Ogino et al. [16] extended Gringarten and Sauty's [12] approach to the problems related to heat extraction from a finite fracture or multiple fractures with a recharge-discharge well pair and obtained good analytical solutions.

The third type is the complete 3D problem for the heat transfer in the fracture and the rock. In these more complicated cases, numerical schemes, such as finite element [7], Marker-and-Cell method [17] and finite difference approaches [18, 19], have to be used to obtain the thermal behavior of the fluid and reservoir. Normally, finding the solution is computationally demanding.

Besides applying the method to a particular set of boundary conditions for EGS reservoirs, the advance of the present model is in that the semi-analytical solutions for the rock and fluid temperatures are obtained, thus significantly reducing computational cost. It also provides an efficient and accurate way to further study the effect of some factors, such as the number of wells, the well spacing, the injection rates and production rates, to maximize the heat extraction from the EGS reservoirs.

2. Problem description

The geometry considered by the present model is shown in Fig. 1. There exist M ($M \geq 1$) vertical recharge well and N-M ($N \geq 2$) vertical discharge wells, which intercepts the fracture containing the injected fluid. The whole system (liquid and rock formation) is initially in an equilibrium state with the uniform temperature T_0. When time $t > 0$, a cold fluid with a constant injection rate Q_i ($i=1, 2...M$) and constant temperature T_{in}^i is injected from the injection points (x_i, y_i) ($i=1,..M$) and the heated fluid is pumped out at a constant production rate Q_i ($i=M+1, M+2...N$), from the discharge wells (x_i, y_i) ($i=M+1, 2...N$).

It is assumed that after a hydraulic fracturing treatment, there is a connected flat fracture plane intercepted by all wells. The geometry of the fracture is defined by its radius In particular, the fracture radius is infinite in this paper.

Some assumptions are made for the present model:

1. The fluid is incompressible with Newtonian rheology and the rock is impermeable, homogenous and isotropic;

2. The stress-induced change of the fracture aperture is ignored;

3. The material properties (density, specific heat capacity and thermal conductivity) of the rock and the fluid are constant and independent of the temperature;

4. The heat condution in the fluid is neglected when the injection rate is sufficiently great.

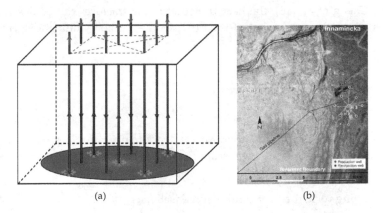

(a)   (b)

Figure 1. (a) Multi-well geothermal reservoir model, (b) Geometry for Habanero project with 5 injection wells and 4 production wells (red denotes production well, blue denotes injection wells).

3. Governing equations

3.1. Pressure diffusion in planar fractures

Because the fluid flow along the narrow fracture is much faster than the reservoir heat temperature changes, it can be assumed to be a pseudo-steady state. This means that the fluid pressure or velocity potential is a function of only the in-plane coordinates x and y. Therfore, the governing equations for the fluid flow obeys the potential flow theory

$$\mathbf{q}_f = -\nabla \phi_f, \tag{1}$$

where ∇ is the gradient operator, $\mathbf{q}_f$ denotes the discharge vector and ϕ_f is the velocity potential for the fluid.

In the present model, there are a total of M recharge wells and N-M production wells to be considered. The continuity equation for the flow of an incompressible Newtonian fluid in the fracture plane is expressed as

$$\nabla \cdot \mathbf{q}_f = \sum_{j=1}^{N} Q_j \delta\left(x - x_j, y - y_j\right), \tag{2}$$

where $\nabla \cdot$ is the divergence operator, δ denotes the Dirac delta function, Q_j is flow rate into or out of the jth well (x_j, y_j) $(Q_j > 0$ for injection and $Q_j < 0$ for production).

3.2. Heat transport in closed fractures

According to Cheng et al. [18], the heat transport in the fractures can be described by the following equation when the effect of the heat storage and longitudinal dispersion can be ignored

$$\rho_w c_w \mathbf{q}_f \cdot \nabla T_f + 2\lambda_r \frac{\partial T_r}{\partial z} = 0, \tag{3}$$

where T_f and T_r denote the fluid temperature and rock temperature, respectively, ρ_w and c_w are the mass density and specific heat, respectively, of the fluid, and λ_r is the thermal conductivity of the rock.

For the surrounding rock, the heat conduction equation is

$$\kappa_r \nabla^2 T_r = \frac{\partial T_r}{\partial t}, \tag{4}$$

where $\kappa_r = \lambda_r / \rho_r c_r$, ρ_r and c_r are the mass density and specific heat, respectively, of the rock formation.

3.3. Initial and boundary condition

The initial temperature distribution for the rock and fluid is assumed to be a constant

$$T_f = T_r = T_0, \tag{5}$$

and the initial velocity potential in the fracture is

$$\phi_f = \phi_0 \quad \text{on } z = 0. \tag{6}$$

It must be mentioned that for the temperature continuity along the fracture, we have

$$T_r = T_f \quad \text{on } z = 0. \tag{7}$$

The temperature at each injection well is fixed

$$T_f = T_{in}^j, \quad \text{at } (x_j, y_j) \ (j = 1,..M), \tag{8}$$

where z denotes the vertical coordinate with $z=0$ being the fracture plane.

In order to avoid the existence of the volume associated with the infinite boundary of the fracture plane, the total injection rate is always equal to the total production rates, which is reasonable for long-term estimatation, i.e.

$$\sum_{j=1}^{N} Q_j = 0. \tag{9}$$

4. Dimensionless formulation

After some manipulation, the above equations can be written in a simplified form

$$\frac{\partial \Theta_r}{\partial \tau} = \frac{\partial^2 \Theta_r}{\partial Z^2},$$

$$\nabla \cdot \mathbf{Q}_f = \sum_{j=1}^{N} \Omega_j \delta \left(X - X_j, Y - Y_j \right),$$

$$\mathbf{Q}_f = -\nabla \Phi_f, \quad \mathbf{Q}_f \cdot \nabla \Theta_f + \chi \frac{\partial \Theta_r}{\partial z} = 0, \text{ on } Z = 0, \tag{10}$$

$$\Theta_f = \Theta_{in}^j, \quad \text{on } (X, Y) = (X_j, Y_j),$$

$$\Theta_f = \Theta_r \quad \text{on } Z = 0,$$

$$\Theta_r = \Theta_f = 0, \quad \text{on } \tau = 0,$$

based on the following transformation

$$\Theta_r = \frac{T_r - T_0}{T_0}, \quad \Theta_f = \frac{T_f - T_0}{T_0}, \quad \Theta_{in}^j = \frac{T_{in}^j - T_0}{T_0}, \quad \Phi_f = \frac{\phi_f - \phi_0}{Q}, \quad \Omega_j = \frac{Q_j}{Q},$$

$$\mathbf{Q}_f = \frac{L}{Q}\mathbf{q}_f, \quad \tau = \frac{t\kappa_r}{L^2}, \quad X = \frac{x}{L}, \quad Y = \frac{y}{L}, \quad Z = \frac{z}{L}, \quad \chi = \frac{2\lambda_r L}{\rho_w c_w Q},$$

where L is the characteristic length to be chosen from the geometrical configuration of the wells (for example, the minimum distance between wells).

5. Velocity potential and stream function

According to the superposition principle, the flow velocity potential Φ for a sum of the injection and producing wells is

$$\Phi = -\frac{1}{4\pi}\sum_{j=1}^{N}\Omega_j \ln\left[(X-X_j)^2 + (Y-Y_j)^2\right], \tag{11}$$

and the flow stream function ψ is expressed as

$$\Psi = -\frac{1}{2\pi}\sum_{j=1}^{N}\Omega_j \arctan\frac{(Y-Y_j)}{(X-X_j)}. \tag{12}$$

The following formulas are valid when the variables Φ and ψ denote the velocity potential function and stream function, respectively,

$$
\begin{aligned}
V_X &= -\frac{\partial\Phi}{\partial X} = -\frac{\partial\Psi}{\partial Y}, & \frac{\partial\Theta_f}{\partial X} &= \frac{\partial\Theta_f}{\partial\Phi}(-V_X) + \frac{\partial\Theta_f}{\partial\Psi}V_Y, \\
V_Y &= -\frac{\partial\Phi}{\partial Y} = \frac{\partial\Psi}{\partial X}, & \frac{\partial\Theta_f}{\partial Y} &= \frac{\partial\Theta_f}{\partial\Phi}(-V_Y) + \frac{\partial\Theta_f}{\partial\Psi}(-V_X).
\end{aligned}
\tag{13}
$$

Then equation (10) $_3$ after using the above identities, becomes

$$-v^2\frac{\partial\Theta_f}{\partial\Phi} + \chi\frac{\partial\Theta_r}{\partial Z} = 0, \quad Z = 0, \tag{14}$$

where

$$v^2 = V_X^2 + V_Y^2.$$

Appling Laplace transform to equation (14) and making use of the general Laplacian solution of equation (10) for Θ_r, the analytical solution for the Laplace tranform of the fluid temperature $\hat{\Theta}_f$ is obtained as

$$\hat{\Theta}_f = \frac{\Theta_{in}}{s}e^{-\chi\sqrt{s}g(\Phi,\Psi)}, \tag{15}$$

where the cap $^\wedge$ denote the Laplace tranform and the function $g(\Phi,\psi)$ is defined as

$$g(\Phi, \Psi) = \int_{\Phi}^{+\infty} \frac{d\zeta}{v^2(\zeta, \Psi)}.$$

The solution for Θ_f in the time domain is obtained by the inverse transformation of Eq. (15)

$$\Theta_f = \Theta_{in} \text{erfc}\left[\frac{\chi g(\Phi, \Psi)}{2\sqrt{\tau}}\right]. \tag{16}$$

We note that in equation (16), Θ_{in} can be measured and χ is a constant. Therefore, the key to the solutions for the temperature is to calculate the value of the function $g(\Phi, \psi)$.

6. Methodology, verification and numerical results

Generally, when the number of the wells is larger than 2, the analytical solutions for $g(\Phi, \psi)$ are difficult and are not available in the literature. Then the following computational procedures are adopted. Start with the points which are very close to one of the injection wells and the velocity, velocity potential Φ_0 and stream function ψ_0 are calculated with the known coordinates. We then move along the streamline, and the velocity potential is increased by an increment $\Delta\Phi$. In addition the coordinates of the next point on the same streamline can be evaluated via the following equations

$$\Psi_0 = \Psi(X,Y),$$
$$\Phi_0 - \Delta\Phi = \Phi(X,Y), \tag{17}$$

which can be solved by using the Newton-Raphson method. In the same way, the information of the next point on the same streamline is obtained until the new point reaches a region which is regarded as being within the range of a production well or as infinity. There are no well defined criteria to define the region which is regarded as being within the range of a production well or as infinity. In the present calculations, we have compared cases using several small radial distances (such as 0.002m, 0.001m and 0.0001m) from the production wells and different far-field distances (such as 6km, 8km and 10km) from all the wells. When we find little difference between the results for these cases, the solutions are taken as being accurate. It should be noted that the position of the new point relative to the total wells should be compared with that of the starting point on the same stream-line in each step. If the relative position changes, the stream function should be plus or minus PI which depends on the position change.

For verification and testing of the proposed model, numerical results for several well arrangements as shown in Fig. 2 are provided below. In particular, the required parameters are listed in Table 1.

Parameters	Value	Parameters	Value
Number of injection wells	M	Rock ther. conductivity λ_r (W/(m•K))	2.2
Number of total wells	N	Rock mass density ρ_r (Kg/m³)	2700
Total injection rate Q_{total} (m³/s)	0.06	Rock specific heat c_r (J/(kg•K))	790
Total Extraction rate Q_{total} (m³/s)	0.06	Fluid mass density ρ_w (Kg/m³)	900
Injection temperature T_{in}	90	Fluid specific heat c_w (J/(kg•K))	4200
Initial temperature T_0 (°C)	270		

Table 1. Parameters for the present calculations

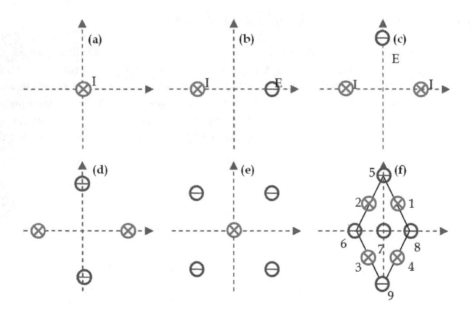

Figure 2. Case studies with different injection wells and extraction wells. Blue circle denotes injection well and red circle denotes production well.

6.1. Cases with one single well and dipole wells

In Figure 3(a), the temperature evolution along the radial direction for the case with one single injection well is compared between the present approach and the analytical solution [5]. In Figure 3(b), the temperature profiles along the line connecting the injection well and the production well are also compared for the case with dipole wells[14] It is found that the results for the above two cases from the current method and analytical solutions are in excellent agreement.

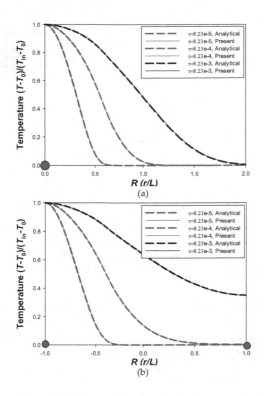

Figure 3. Comparison between the present approach and the analytical solutions: (a) Temperature profile along the radial direction (physical time *t*=115 days, 3.17 years and 31.7 years) for the case with only a single injection well at the origin. Parameters are listed in the paper by Ghassemi et al. [6]; (b) Temperature distribtion along the line connecting a recharge well (left) and a discharge well (right). Blue circle denotes injection well and red circle denotes extraction well.

6.2. Cases with two injection wells or multiple wells

Figure 4 displays the extraction temperature change at different times for the cases (b) to (e) (Figure 2). As the geometrical configurations of all cases are symmetrical, the output temperature for only one production well in each case is plotted. The total injection and production rates in all cases are kept the same, i.e. 0.06 m³/s, and for the purpose of simplicity, the individual injection rate for each case is assumed to be the same. For example, when there are two injection/production wells, the individual injection/production rate is 0.06/2=0.03 m³/s. These results show that case (b) has the least temperature decrease, the reason being that the well separation is greater than the other cases. However, this greater separation and the fact that only 2 wells are involved will undoubtedly lead to higher pressure differences in order to obtain the same total flow rate. Case (d), with 2 injectors and 2 producers, is the next best performing well from the standpoint of minimizing temperature falloff. Interestingly, case (c) and case (e) perform almost equally, but case (c) only requires 3 wells while case (e) requires 5.

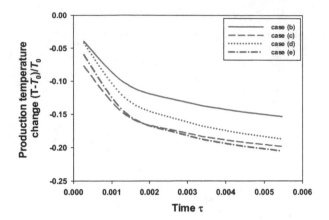

Figure 4. Normalized change in production temperature from cases (b) to (e) under the same total injection and production rates $Q=0.06 m^3/s$. The average value of the temperatures at all the points on the small circle around the production well is chosen as the production temperature.

Figure 5 shows the output temperature change for different geometrical sizes and injection rates are plotted for case (c) (Figure 2). $X=2\lambda_r L/(\rho_w c_w Q)$ is an important parameter, where L is the well separation and Q is the total flow rate. When the value of χ is kept constant, the fluid temperature at the output will evolve equivalently.

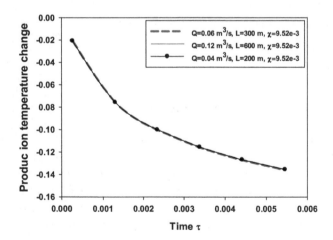

Figure 5. Normalized temperature change for the production wells while keeping the controlling parameter set to $\chi=4.76e-3$.

Figure 6 presents isothermal lines at 5 years and 21 years for case (d). We can see clearly the contours at the same temperature propagating away from the injection well with time. As there are associated with the singularity in flow velocity at the production well whereby the temperature there cannot be obtained analytically and is calculated approximately. In order to make this approximation, a small circle around the well is chosen and calculations are not carried out inside this region. The number of the streamlines used in the computation which flows into this well can be counted and the temperature averaged over all used streamlines can also be computed and chosen as the production well temperature. In Figure 6, because of the symmetry, each production well accepts flow from each injection well. However, which injection well the influx at a specific part of a well comes from depends on the angle at which it faces toward an injection well. Therefore, the temperature fields around the production wells are very complicated and require an averaging method.

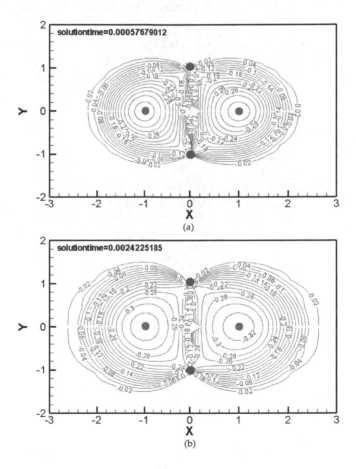

Figure 6. Isothermal lines of the normalized temperature change at (a) time=5 years (τ=5.77e-4) and (b) time=21 years (τ=2.42e-3) for case (d). Blue circle denotes injection well and red circle denotes production well.

The streamline profiles and production temperature in Figure 7 are for case (f) (Figure 2) where there are four injection wells and five production wells. The total flow rate is also 0.06m³/s.

Due to the symmetry, the temperature distributions of injection wells 1, 2, 3 and 4 are similar, as are the temperature fields between wells 5 and 9 and between production wells 6 and 8. Figure 7(b) shows the trend in the average normalized temperature variations in time. On the other hand, we can find that the contribution to the normalized temperature change of a particular well may strongly depend on the well arrangement, as well as on the imposed outflow rate at the production well, based on the temperature changes in Figure 7(b). For instance, all the four injection wells contribute to outflux at well 7 (the middle one in case (f)), but its output temperature undergoes the biggest decrease as show Figure 7(b). This is attributed to a larger flow rate in this central region, defined by four injection wells. The heat exchange from the rock to the fluid is not sufficient to heat the fluid to the same degree as

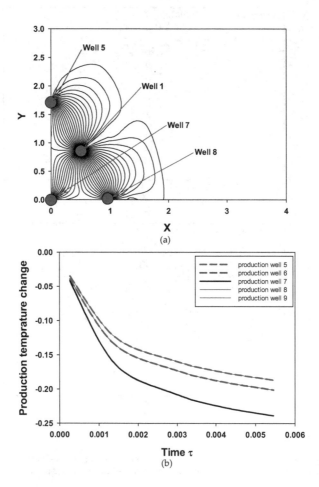

Figure 7. (a) Streamlines profiles and (b) normalized production temperature change in case (6) when $Q_{in}=Q/4=0.015$ m³/s and $Q_{out}=Q/5=0.012$ m³/s. Plot (a) shows a quarter of the whole streamline profiles which are symmetric across the plot axes.

occurs at other production wells. In the end, this will affect the output temperature at well 7. It must be remembered that all production wells are assumed to have the same outflux. Therefore, some fluid heated in the central region is forced to move outward and eventually reaches the outer production wells. The direct result of the heated fluid movement is to increase the output temperature at the outer production wells.

7. Conclusions

By using the Laplace transformation, the analytical form of the solutions in the Laplace space and thus, by inversion, in the time domain are obtained. The approach provides an efficient and accurate way to calculate the rock and fluid temperatures. Through several preliminary case studies, some brief conclusions are obtained:

1. Even for a multi-well geothermal reservoir, we obtain analytical or semi-analytical solutions that provide an efficient and accurate way for predicting the rock and fluid temperature;

2. The flow rates and the relative locations of the wells determine the flow path of the fluid. The injection rate also determines the thermal behavoir of the fluid and the parammeter $X=2\lambda_r L/(\rho_w c_w Q)$ reflects the rate of heat extracted by the cold injection fluid.

3. The efficiency of heat extraction from the EGS reservoir studied here depends on the layout and spacing of the injection and production wells. In fact, our model showed some contrasting cases where fewer wells outperformed a greater number of wells. Ongoing work will be aimed at finding optimal well layouts and flow rates.

Author details

Bisheng Wu[1*], Xi Zhang[1], Andrew Bunger[1,2] and Rob Jeffrey[1]

*Address all correspondence to: bisheng.wu@csiro.au

1 CSIRO Earth Science and Resource Engineering, Melbourne, Australia

2 Department of Civil and Environmental Engineering, University of Pittsburgh, Pittsburgh, PA, USA

References

[1] Lauwerier, H. A. The transport of heat in an oil layer caused by the injection of hot fluid, Applied Scientific Research, (1955). , 5, 145-150.

[2] Bödvarsson, G. S. On the Temperature of Water Flowing through Fractures. Journal of Geophysical Research, (1969). ,74(8),1987-1992.

[3] Gringarten, A. C. Theory of heat extraction from fractured hot dry rock, Journal of Geophysical Research, (1975).

[4] Zhang, X, Jeffrey, R. G, & Wu, B. Two basic problems for hot dry rock reservoir stimulation and production. Australian Geothermal Conference (2009). Brisbane, QLD, Australia, Nov., 2009., 11-13.

[5] Mossop, A. Seismicity, subsidence and strain at the Geysers geothermal field. PhD. Dissertation, Stanford University, (2001).

[6] Ghassemi, A, Tarasovs, S, & Cheng, A. H. D. An integral equation solution for three-dimensional heat extraction from planar fracture in hot dry rock. International Journal for Numerical and Analytical Methods in Geomechanics, (2003). , 27, 989-1004.

[7] Kolditz, O. Modelling flow and heat transfer in fractured rocks: conceptual model of a 3-D deterministic fracture network. Geothermics, (1995). , 24(3),421-437.

[8] Gringarten, A. C, & Ramey, H. J. The use of source and Green's functions in solving unsteady-flow problems in reservoirs. Society of Petroleum Engineers Journal, (1973). , 285-296.

[9] DaCosta, J.A, & Bennett R.R., The pattern of flow in the vicinity of a recharging and discharging pair of wells in an aquifer having areal parallel flow. International Association of Scientific Hydrology, (1960). , 52, 524-536.

[10] Grove, D. B, Beetem, W. A, & Sower, F. B. Fluid travel time between a recharging and discharging well pair in an aquifer having a uniform regional flow field, Water Resources Research,(1970)., 6(5), 1404-1410.

[11] Muskat, M. The flow of homogeneous fluids through porous media. MicGraw-Hill Book Company, Inc., (1937).

[12] Gringarten, A. C, & Sauty, J. P. A Theoretical Study of Heat Extraction from aquifers with uniform regional flow. J. Geophys. Res., (1975)., 80(35),4956-4962.

[13] Rodemann, H. Analytical model calculations on heat exchange in a fracture. Urach Geothermal Project, Haenel R. (ed.), Stuttgart, (1982).

[14] Schulz, R. Analytical model calculations for heat exchange in a confined aquifer. J. Geophys. Int., (1997). , 61, 12-20.

[15] Heuer, N, Kopper, T, & Windelberg, D. Mathematical model of a Hot Dry Rock system. Geophys. J. Int., (1991). , 105, 659-664.

[16] Ogino, F, Yamamura, M, & Fukuda, T. Heat transfer from hot dry rock to water flowing through a circular fracture. Geothermics, (1999). , 28, 21-44.

[17] Harlow, F. H, & Pracht, W. Theoretical study of geothermal energy extraction. Journal of Geophysical Research, (1972).,77(35), 7038-7048.

[18] Cheng, A. H. D, Ghassemi, A, & Detournay, E. Integral equation solution of heat extraction from a fracture in hot dry rock. International Journal for Numerical and Analytical Method in Geomechanics, (2001)., 25, 1327.

[19] Ghassemi, A, Tarasovs, S, & Cheng, A. H. D. Integral equation solution of heat extraction-induced thermal stress in enhanced geothermal reservoirs, International Journal for numerical and Analytical Methods in Geomechanics, (2005). , 29, 829-844.

Chapter 7

Scale Model Simulation of Hydraulic Fracturing for EGS Reservoir Creation Using a Heated True-Triaxial Apparatus

Luke Frash, Marte Gutierrez and Jesse Hampton

Additional information is available at the end of the chapter

Abstract

Geothermal energy technology has successfully provided a means of generating stable base load electricity for many years. However, implementation has been spatially limited to rare high quality traditional resources possessing the combination of a shallow high heat flow anomaly and an aquifer with sufficient permeability and fluid recharge. Enhanced Geothermal Systems (EGS) technology has been proposed as a potential solution to enable additional energy production from the much more common non-traditional resources. To advance this technology development, a heated true triaxial load cell with a high pressure fluid injection system has been developed to simulate an EGS system from stimulation to production. This apparatus is capable of loading a 30x30x30 cm3 rock sample with independent principal stresses up to 13 MPa while simultaneously providing heating up to 180 °C. Multiple orientated boreholes of 5 to 10 mm diameter may be drilled into the sample while at reservoir conditions. This allows for simulation of borehole damage as well as injector-producer schemes. Dual 70 MPa syringe pumps set to flow rates between 10 nL/min and 60 mL/min injecting into a partially cased borehole allow for fully contained fracturing treatments. A six sensor acoustic emission (AE) array is used for geometric fracture location estimation during intercept borehole drilling operations. Hydraulic pressure sensors and a thermocouple array allow for additional monitoring and data collection as relevant to computer model validation as well as field test comparisons. The results of the scale model hydraulic fracturing tests demonstrate the functionality of the equipment while also providing some novel data on the propagation and flow characteristics of hydraulic fractures. Fully characterized test sample materials used in the scale model tests include generic cement grout, custom high performance concrete, granite, and acrylic. Fracturing fluids used include water, brine, and Valvoline® DuraBlend® SAE 80W90 oil.

Keywords: Enhanced Geothermal Systems (EGS), true triaxial device, hydraulic fracturing, scale model testing, acoustic emissions

1. Introduction

The potential of Enhanced Geothermal Systems (EGS) is well documented in the MIT led study titled "The Future of Geothermal Energy" [1]. With this technology, unconventional deep Hot Dry Rock (HDR) reservoirs are engineered with drilling and stimulation techniques to create a heat mining system for base load energy production. The methods needed for enabling EGS energy production also have the ability to improve production from traditional geothermal resources which are already being utilized today.

To provide the EGS reservoir stimulation, one of the most promising techniques is hydraulic fracturing. This method utilizes high pressure fluid injection into targeted reservoir intervals to enhance permeability and generate new flow paths through enhancing existing fractures and creating new fractures. With the installation of an injector-producer well scheme, the physical limitations of natural reservoir recharge and stored harvestable fluids may be overcome and a productive reservoir may be the end result. Hydraulic fracturing has been proven effective as a stimulation technique by the oil and gas industry since its first implementation in 1947 [2].

Currently, only a small number of EGS field trials have been performed due to the high economic risk of the procedure and the significant probability of failure. Thus, performing controlled EGS experiments in the laboratory setting may be able to provide some of the crucial data and experience needed for advanced fracture model calibration and full scale testing in the field. This is especially true considering that most hydraulic fracturing design techniques, as developed by the petroleum industry, are more dependent upon historical data than on theoretical analysis [3]. In the case of EGS development, this historical data does not yet exist in sufficient quantities.

To fill the knowledge gap, laboratory scale EGS reservoir testing is being performed at the Colorado School of Mines using a heated true-triaxial apparatus. Some completed test results and observations are presented along with technical information on the equipment and procedures used. Focus is given to series of tests performed on a hydraulically fractured granite sample with a binary injector-producer borehole scheme installed.

2. Equipment design and specifications

The laboratory scale EGS simulation equipment consists of four main subsystems being a heated true-triaxial cell, a high pressure hydraulic injection system, a multi-component data acquisition system, and sample materials and characterization equipment.

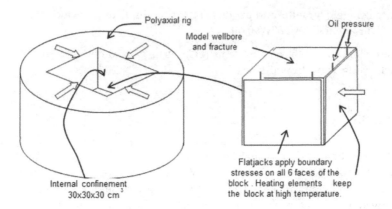

Figure 1. Layout of the true-triaxial cell.

2.1. Heated true triaxial cell

The layout of the heated true triaxial cell is shown in Figure 1. It consists of a cylindrical loading rig made of high strength steel. Flatjacks apply pressures on all six faces of a 30x30x30 cm³ block rock sample. Freyssinet 350 mm flatjacks, which are pressurized with pumps, allow independent control of the principal stresses of up to 12.5 MPa. The flatjack pressures can be controlled to achieve triaxial stress conditions with different magnitudes of overburden stress σ_v, maximum horizontal stress σ_H, and minimum horizontal stress σ_h. Externally mounted flexible silicone rubber heaters with proportional-integral-derivative (PID) control allow for dual-zone heating with separate set points for the lateral and vertical heating elements. The heating system allows for the simulation of an EGS reservoir with a temperature of up to 180 °C.

Figure 2 shows pictures of the completed true triaxial cell with and without the drilling rig placed on top of the cell. An orientated rotary-hammer drill press is used to drill boreholes into the sample at user selected positions and angles while the sample is under stresses and temperature. This procedure allows for strategic borehole installations that are specific to the test and the particular stimulated fracturing plane. Borehole damage is replicated by using percussive drilling into the loaded sample instead of the more common cast-in-place pre-drilled borehole methods[4-7]. The borehole is typically drilled with one upper cased segment having a maximum outside diameter of 10 mm and a second uncased fracturing interval having a typical diameter of 5.6 mm. These dimensions were selected to be as small as possible to allow for the most effective EGS reservoir simulation within the confines of the 30x30x30 cm³ cubical sample blocks.

2.2. High pressure hydraulic injection system

A programmable hydraulic injection system is used for both hydraulic fracture stimulation and post-fracture flow analysis. Precision high pressure flow is provided by a dual 65DM Teledyne Isco syringe pump system, a series of pneumatic-hydraulic automated valves, and a custom pump control program developed with LabVIEW. This system is capable of providing pressures up to 70 MPa and precise controlled flow rates between 10 nL/min and 60

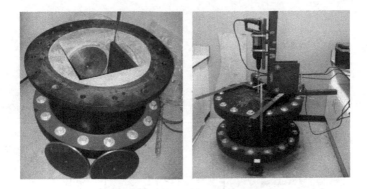

Figure 2. Pictures of the true triaxial cell. Left: without the drilling rig, Right: with the lid and drilling rig.

mL/min with a flow stability of ±0.3% from the set point. A diagram of the hydraulic system is provided in Figure 3.

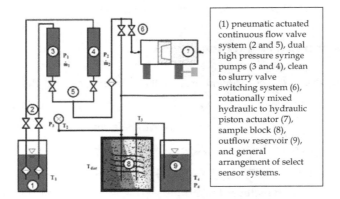

(1) pneumatic actuated continuous flow valve system (2 and 5), dual high pressure syringe pumps (3 and 4), clean to slurry valve switching system (6), rotationally mixed hydraulic to hydraulic piston actuator (7), sample block (8), outflow reservoir (9), and general arrangement of select sensor systems.

Figure 3. Diagram for the hydraulic fracturing system

Some of the programmable capabilities of the system include: (1) Stepwise continuous constant flow or pressure, (2) Controlled switching between clean and slurry fluid injection, and (3) Conditionally dependent operation with real time external data referencing capability. To seal the injection tubing into the borehole, threaded 316 SS tubing was grouted into a 10 mm outside diameter borehole using Loctite® Rapid Mix 5-Minute epoxy. The epoxy grout was delivered downhole using water-softened 00-size gelatin capsules to avoid the potential of bonding the casing to the true-triaxial cell's top lid. After reaching a 24 hr cure, an uncased 5.6 mm diameter interval was drilled through the bottom of the casing and into the sample. Figure 4 shows a diagram of the borehole sealing method.

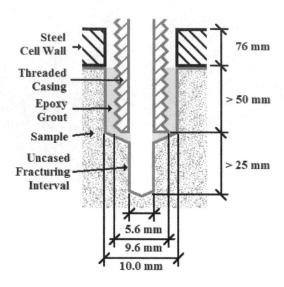

Figure 4. Borehole sealing method applied with typical dimensions.

2.3. Multi-component data acquisition system

To monitor and control the equipment and system processes a multi-channel multi-function National Instruments CompactDAQ was used with 16 strain gage channels, 16 CJC thermo-couple channels, 8 voltage channels, 8 current channels, and 4 multi-function channels. The attached sensors included 2 Omega® PX309-10KG5V pressure transducers for monitoring the injection wellhead pressure and intermediate principal sample confining stress, 1 Omega® PX309-3KG5V pressure transducer for monitoring the minimum principal stress, 1 Omega® PX40-50mmHG pressure transducer for monitoring the production reservoir fill level and flow rate, 1 Omega® LD621-30 linear displacement transducer for auxiliary use, and 1 Humboldt HM2310.04 linear strain transducer also for auxiliary use. Omega® Type-T thermocouples, fabricated in-house, were positioned at the hydraulic temperature monitoring positions as indicated in Figure 3, inside the bottom of the injection and production boreholes, and in a high-coverage grid arrangement on the surface faces of the sample inside the cell, as shown in Figure 5. When used, strain gages were embedded onto the faces of the sample to monitor stress uniformity. Additional data was collected from the Teledyne Isco pump controller giving information about hydraulic system operation including flow rates, pressures, valve positions, and general pump status.

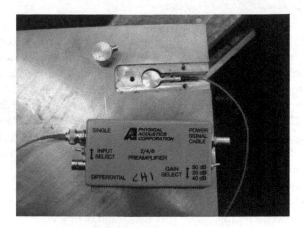

Figure 6. AE sensor installed in a loading platen.

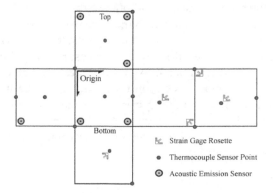

Figure 5. Diagram of surface sensor positions on a typical 30x30x30 cm³ sample.

To monitor the fracturing process and provide real-time location estimation for the generated hydraulic fractures, a 6-sensor piezoelectric Acoustic Emission (AE) monitoring system, obtained from Physical Acoustics Corporation, was installed inside the cell with sensors contacting the faces of the sample in an arrangement to achieve maximum volumetric coverage, as shown in Figure 5. Figure 6 shows an AE sensor installed into a 25 mm thick loading platen where it was protected from the high loading stresses being applied to the sample. Thin packing foam wafers were inserted between the sensor body and the steel housing to dampen external acoustic noise effects and provide a soft spring reaction for any movement that would occur during loading and unloading processes. In general, this platen serves as a movable interface between the pressurized flat jack and the sample inside the cell. During analysis, recorded AE events could readily be filtered by correlation coefficient, amplitude, or other criteria using digital post-processing of hit time and waveform data.

2.4. Test materials

Four material types were used for this project including medium strength concrete grout, ultra-high strength low permeability concrete, locally obtained Colorado Rose Red Granite, and acrylic glass. Each of these materials was tested for a variety of mechanical, thermal, and acoustic properties to provide a reference for future field data comparison. A general summary of the measured properties for selected materials has been provided in Table 1. The uniaxial compression strength (UCS), elastic modulus (E), Poisson's Ratio (v), and indirect tensile strength (BTS) testing was performed using a specially instrumented ELE Accu-Tek™ 250 concrete load frame. Thermal conductivity (k_T) measurements were performed using a divided bar apparatus available through the Colorado Geological Survey [8]. Volumetric specific heat capacity (C_V) was obtained using an insulated calorimeter. Acoustic compression (V_P) and shear (V_S) wave velocities were obtained using a piezoelectric pulse transmitter-receiver apparatus with oscilloscope monitoring. Porosity (ϕ) and matrix density (ϱ_{dry}) were measured using a 70% vacuum desiccator, 110ºC oven, and digital mass balance.

For post-test analysis, diamond over cores and cut cross-sections were used. The over cores were taken to remove the borehole casing and observe the near wellbore fracture geometry. Next, cross-sections were cut using a 0.9 m diameter diamond table saw. An example cross section taken from an unconfined granite sample hydraulic fracturing test is shown in Figure 7. Cross sections such as these allowed for physical measurements of the fracture locations, fluid permeation depths, and verification of AE fracture location estimations. Fluid pathways and permeation depths were most visible on tests using oil as the fracturing fluid due to staining of the sample material. Compiling fracture geometry data from consecutive cross sections allows for three-dimensional imaging of entire stimulated fracture networks. As evident in Figure 7, these networks are expected to be very complex due to the heterogeneities in natural rock and concrete samples.

3. Test results and observations

Using this equipment, an ongoing series of hydraulic fracturing stimulation and reservoir characterization testing is being performed to obtain new data for EGS technology advancement. While hydraulic fracturing experiments have been performed in more than 11 different boreholes and four different materials, focus will be given to a granite hydraulic fracturing test where an orientated intercept borehole was drilled to create a producing heated EGS reservoir. The results of the EGS simulation experiment can be divided into several key phases including sample preparation, primary hydraulic fracturing, drilling the fracture intercept borehole, and fracture reopening and flow.

3.1. Sample preparation

For this test, a block of Colorado Rose Red Granite, as documented in Figure 8, was loaded into the true-triaxial cell and slowly heated to an average internal temperature of 50 ºC over the span

Material Property	Medium Strength Concrete	Ultra-High Strength Concrete	Colorado Rose Red Granite
Unconfined Compressive Strength, UCS (MPa)	50-60	123-154	152 ± 19[*]
Brazilian Tensile Strength, BTS (MPa)	2.2-2.7	4.0-6.0	7.5 ± 1.8[*]
Young's Modulus, E (GPa)	9.5-10.5	20-30	57[*]
Poisson's ratio, v	-	-	0.32[*]
Dry density, ρ_{dry} (kg/m³)	1950	1970	2650
Thermal conductivity, k_T (W/m-K)	-	1.60 ± 0.02	3.15 ± 0.05
Heat capacity, C_V (kJ/m³-K)	2013 ± 145	1820 ± 146	2063 ± 92
Porossity, ϕ	0.30-0.31	0.15-0.23	0.006-0.008
Shear wave velocity, V_S (mm/µs)	2.48	2.54	2.62
Compressional wave velocity, V_P (mm/µs)	3.41	3.89	4.45

Table 1. Test Material Properties (*data from [9-10]).

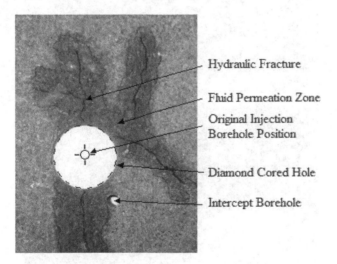

Figure 7. Cross-section from an unconfined granite hydraulic fracturing test.

of four days. After the target temperature was reached, the sample was pressurized with confining stresses of 12.5, 8.3, and 4.1 MPa for the vertical, maximum horizontal, and minimum horizontal stresses, respectively. The AE monitoring system was active throughout the loading process to identify if any mechanical shearing or thermal fracturing events had occurred. In this case, the AE data produced a large scatter of events with no significant clustering which indicated that acceptably uniform loading had been achieved and no significant fracturing events had occurred. The uniformity of the sample loading was also verified using strain gage data, with the layout as shown in Figure 5. Combined, both of these methods were in agreement that a top corner of the sample was subjected to some elevated stress concentrations as indicated by relatively high strains and an increase in localized AE activity at the specific corner. This observation was used to modify and improve the loading procedure such that similar unintended stress concentrations would be less likely to occur during future tests using this equipment.

While loaded, a centered vertical borehole was drilled into the sample, a 107 mm deep casing interval was installed, and a 73 mm uncased interval was drilled for a final injection well depth of 180 mm. It is important to note that drilling the borehole while the sample is under load is a unique system capability that allows for laboratory simulation of a borehole damage zone. This process creates small fractures near the borehole, as has been clearly observed in acrylic testing [11-12], which may serve as fracture initiation locations. Simultaneously, the drilling process also fills these micro fractures with fines which are believed to have some effect on fracture self propping as well as near wellbore tortuosity and skin factor. Additional investigation may be necessary to better understand how the borehole damage zone influences hydraulic fracture initiation, growth and closure.

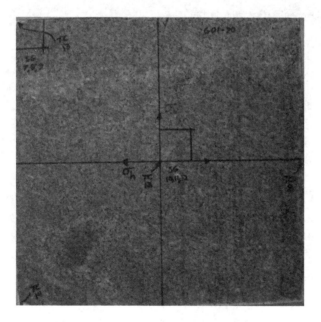

Figure 8. Pre-test image of the granite sample used for EGS reservoir simulation experiments.

3.2. Primary hydraulic fracturing

Primary hydraulic fracture breakdown was achieved using oil injection at a constant flow rate of 0.05 mL/min. Valvoline® Durablend® SAE 80W90 gear oil was used as the fracturing fluid due to its high viscosity value and publicly available fluid properties. At the injection temperature of 50 ºC, this fluid has an approximate dynamic viscosity of 71.5 cP as estimated using the published product information in conjunction with the Walther Equation specified in ASTM D341 [13]. The importance of using high viscosity fluid for laboratory hydraulic fracture experiments is well documented [5-6]. In this case, using a high viscosity fluid provided the important benefits of better fracture growth control for improved probability of containment and a more predictable fracture orientation as the propagation would be less influenced by natural heterogeneities in the granite sample.

A plot of the hydraulic data for primary breakdown is shown in Figure 9. During this test, the pump was stopped 16 seconds after breakdown in an attempt to keep the fracture fully contained as real-time AE events were observed to be approaching the edges of the sample. Continued AE activity was observed even after pumping was stopped which indicated continued fracture propagation. Therefore, to forcibly halt the fracture growth, the flow rate was reversed at -10 mL/min for a total of 6 seconds to pull fluid out of the fracture and then held in the stopped position thereafter. At this time, a significant pressure rebound was observed which may offer some insight into fracture dynamic fluid storage behavior with additional investigation. Ultimately, the observation of a negligible flow rate during post-fracture constant pressure testing at 2000 kPa verified that a fully contained fracture had been generated.

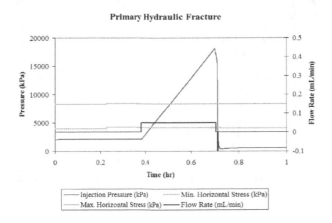

Figure 9. Hydraulic data plot for primary fracture.

Analysis of the AE source location data collected during this primary hydraulic fracture test revealed that a contained and planar fracture propagated from the borehole in a direction perpendicular to the minimum horizontal confining stress. Additionally, the fracture appeared to have a single dominant wing as evident by the AE cloud being most prominent on only one side of the borehole. Figure 10 shows orthogonal plots of the three-dimensional AE event source location results for the test. This analysis used six-sensor location regression and filtered the results to only contain events with a correlation coefficient greater than 0.75 and amplitude greater than 25 dB. On this plot, the circle diameters are directly proportional to the amplitude of the corresponding event. Also, the color shading corresponds to the correlation coefficient

AE Event Category	Number	% Total	% Classification
Total Events Located	726	100	-
Classifiable Events	81	11.2	100
Tensile Events	39	5.4	48.1
Shear Events	28	3.9	34.6
Mixed Mode Events	14	1.9	17.3

Table 2. Classifications of AE Events during hydraulic fracturing.

of each event with dark red circles having higher correlation. The two-segment centered vertical injection borehole is clearly visible on the front and side view plots.

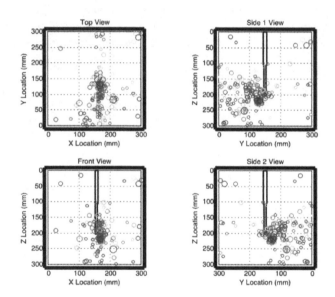

Figure 10. AE event source locations during primary hydraulic fracture.

Extending the AE analysis by application of moment tensor methods [14], information was obtained about the fracturing mode for some of the recorded AE events. As shown in Table 2, only about 11% of the total number of recorded events could successfully be classified with a reasonable level of certainty. At a glance, the tensile failure mode appears to be dominant during this fracturing stage but uncertainty associated with the low percentage of classifiable events effectively reduces the confidence of any conclusions which could possibly be derived from these figures.

3.3. Drilling the fracture intercept production borehole

Using AE source location data, an estimate of the fracture geometry was obtained and an optimal intercept borehole position was selected as shown in Figure 11. Here, the intercept borehole trajectory, drilled at 30° from the vertical axis, can be seen penetrating through the expected fracture surface. A high-angle drilling orientation was used to maximize the probability of achieving a successful intercept after considering AE source location uncertainty and drilling system tolerances. Also, the uncased 10 mm diameter intercept borehole was drilled deeper than the expected intercept location to further increase the probability of successful hydraulic connection. In the figure, the best estimate of the fracture plane was plotted using a smoothed cubic interpolation surface function fitted to events with both high-amplitude and high-correlation. After drilling was completed, the borehole was swabbed and positive indication of fracturing oil was recovered, thus indicating that the intercept was successful.

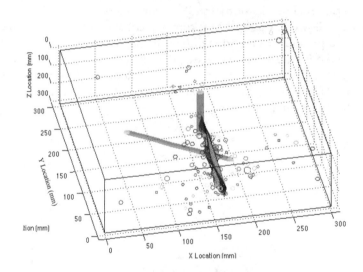

Figure 11. AE generated fracture surface of initial hydraulic fracture.

3.4. Fracture reopening and circulation flow

With the completion of the simulated EGS reservoir, flow experiments were performed to characterize the hydraulic properties of the reservoir. These experiments included constant pressure steady state injection, constant flow rate injection for fracture reopening, stepped constant pressure injection, and constant flow rate injection without reopening. The results obtained from these tests ultimately verified that a hydraulic circuit was present inside of the sample connecting the injection borehole to the production borehole through the stimulated hydraulic fracture.

Initially, constant low-pressure steady-state SAE 80W90 oil injection was performed using specified pressures of 2000, 3000 or 4000 kPa. The pressures were intentionally kept below the minimum principal stress to avoid the potential for continued fracture propagation which could occur with fracture reopening. The results from these tests demonstrated that the achievable stable flow rates with the primary hydraulic fracture geometry were negligible and thus the reservoir remained non-producing. While this information confirmed that the stimulated fracture geometry was fully contained as desired, it also indicated that the connection between the injection and production boreholes was too tight to pass any significant amount of fluid through. It is expected that a significantly higher post-fracture hydraulic conductivity would occur if proppant had been used during the primary fracturing stage.

To enhance the hydraulic connection of the binary borehole system, two fracture reopening stages were performed with stepped constant pressure injection tests executed in between for diagnostic purposes. These injection tests continued to use oil as the injection fluid as its high viscosity was favorable for generating controlled fractures. Figures 12 and 13 show plots of the hydraulic data obtained from the first and second fracture reopening stages respectively. Both of these plots clearly show classic hydraulic fracture reopening behavior [15] with a nearly linear pressure rise followed by a rapid breakdown event and pseudo-steady fracture propagation at an elevated pressure. Comparing the similar magnitude peak pressures of 18.1, 15.4, and 17.4 kPa, observed for the primary fracture, first reopening, and second reopening events respectively, suggests that fracture toughness was not a dominant factor in fracture propagation so scaling criterion suggested in the literature (e.g. [5]) are likely to be satisfied even with intact granite as the testing material.

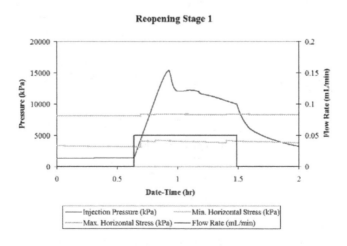

Figure 12. First fracture reopening event.

An orientated view of the AE source location data as observed for the first reopening stage is shown in Figure 14. Comparing this figure to the results shown in Figure 10 and the data from the second reopening stage, it is apparent that most of the fracture growth occurred during the first reopening stage along the bottom and two horizontal extremities of the initial fracture plane. Additionally, the close proximity of the AE events to the boundaries of the sample suggested that the stimulated fracture may no longer have been fully contained and lower fluid recovery efficiency during production could result. The propagation of the fracture to the sample boundary, while not ideal, was reminiscent of the extension of a hydraulic fracture into a faulted zone or natural high-flow fracture network. Here the relative permeability between the sample boundary and the cell's platens was expected to be much higher than that through the hydraulic fracture within the sample, just as a faulted zone would likely have a higher permeability than an artificially stimulated fracture. This situation, while not ideal, may more closely resemble high fluid loss field EGS systems such as those encountered at Hijiori, Japan, where treatments were performed within a discontinuous and naturally fractured volcanic zone [16]. For the final fracture geometry within the granite sample, as estimated with the AE source location data, the smaller wing of the initial fracture appeared to have extended to approximately match the dominant wing length, thus creating a planar bi-wing fracture.

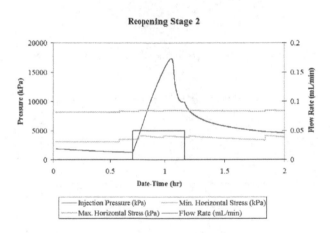

Figure 13. First fracture reopening event.

Comparing the AE count frequency data with the pressure data, as shown in Figure 15, significant increases in AE activity were found to occur just after portions of the hydraulic data where the second derivative of injection pressure with time was negative. Thus, from observing the real-time rate of slope change in the pressure data, it may be possible to anticipate a major fracture growth event before it occurs. Also, using a technique such as this allows for an improved understanding of fracture growth behavior in heterogeneous systems during the time between fracture initiation and shut-in. During this time, the second-order analysis could be used to identify distinct breakdown events occurring after the initial breakdown as could be expected with multi-wing fracture systems or the opening of intersected fissures, joints, or fault zones. In this laboratory case, the analysis was performed using an 11-second backward linear regression approach to obtain an estimate of the first pressure derivative, as could be used in real-time applications. The 11 second value was selected using a qualitative trial-and-error approach with the goal of obtaining a visually smoothed data set without sacrificing too much of the data accuracy.

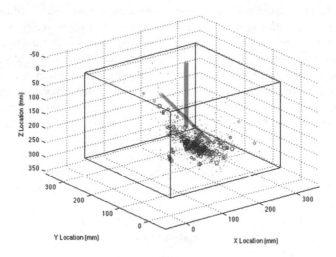

Figure 14. Three dimensional view of AE event source locations during first fracture reopening stage.

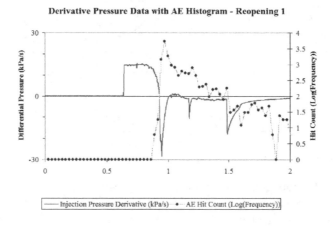

Figure 15. First time derivative of pressure with AE hit count histogram for the first fracture reopening stage.

To evaluate the effectiveness of each fracture reopening stage, stepped constant pressure oil injection tests were performed. In these tests, SAE 80W90 oil was injected into the sample with PID controlled pressure at 1000 kPa increments with 30 minute duration. An example of the hydraulic data from a step pressure test performed after the second reopening event is provided in Figure 16. For each constant pressure increment, the resulting steady state pressure and flow rate measurements are averaged to estimate the pressure dependent flow characteristics of the stimulated reservoir. These values were useful reference points during later controlled constant flow tests where fracture reopening and extension pressures were not desired.

A comparison of the stepped constant pressure test data obtained before and after the second reopening event is shown in Figure 17. On this plot, it was evident that there was negligible flow rate dependence with pressure after the first fracture reopening stage. This suggested that the flow of the injected fluid was not dominated by stimulated fracture flow and the hydraulic connection between the injection and production boreholes was not flowing effectively if at all. To improve the inter-well connectivity, the second fracture reopening stage was performed with high success. As can be seen in Figure 17, pressure dependent flow rate characteristics were much more prominent after this second stage with a clear proportional relationship. To augment these observations, borehole swabbing was performed periodically to check for fluid production in the intercept borehole. The swab's results did not positively indicate hydraulic connection until after the second fracture reopening stage. Thus, even though the first treatment did not attain an acceptable hydraulic connection, the execution of additional fracture stimulation treatments from the same injection well was successful in creating an effective hydraulic connection

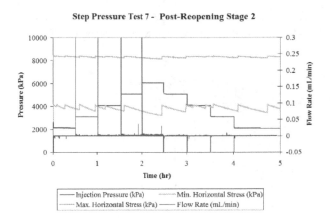

Figure 16. Step pressure test data taken after the second fracture reopening stage.

With a confirmed hydraulic connection between the boreholes, the injection fluid was changed to tap water for thermal flow testing and EGS reservoir characterization. Water injection was performed with two constant flow rate controlled tests to attain pressure dependent flow characteristics for the reservoir. The first test utilized a flow rate of 0.05 mL/min and the second used a higher rate of 0.10 mL/min. While these tests provide similar data to constant pressure injection, it is more easily compared to field applications where flow rate control is the standard. Periodic borehole swabbing results indicated a significant and continuous fluid production in the intercept borehole. Figure 18 provides an example of the flow rate data obtained during the second water flow test. Here, it is evident that the reduction in viscosity by changing from oil to water resulted in significantly reduced pressure losses, as expected. Also, these flow rates did not produce any significant AE activity indicating that the stimulated

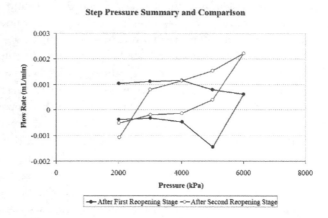

Figure 17. Averaged oil injection stepped constant pressure data before and after the second fracture reopening stage.

fracture geometry was stable with water flow. Additional testing is required and planned in order to obtain a full characterization of the laboratory simulated EGS reservoir.

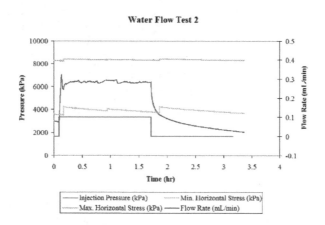

Figure 18. Second water flow test data plot.

4. Conclusions

A heated true-triaxial cell has successfully been able to produce a laboratory simulation of an EGS reservoir. Preliminary experiments using granite have provided valuable resulting data as well as new observations that may bring some additional insight into the potential of EGS technology. Some of the most important advancements and observations that have been made include:

- The completed development of a heated true-triaxial cell with the ability to simulate multi-well EGS reservoir systems as well as borehole damage by percussively drilling orientated boreholes into a hot stressed sample.

- The successful laboratory simulation of a binary injector-producer EGS reservoir in granite with proven fluid communication through a stimulated fracture between two boreholes.

- Multiple hydraulic fracture stimulation treatments may be performed from the same injection borehole to attain significantly increased reservoir conductivity and well fluid communication.

- Significant fracture growth, as indicated by AE activity, is preceded by periods where the real-time second order differential of pressure with time is negative.

- AE source location is a functional and important tool for successful drilling of a production well into a stimulated EGS reservoir.

Acknowledgements

Financial support provided by the U.S. Department of Energy under DOE Grant No. DE-FE0002760 is gratefully acknowledged. The opinions expressed in this paper are those of the authors and not the DOE.

Author details

Luke Frash, Marte Gutierrez and Jesse Hampton

Colorado School of Mines, Golden, CO, USA

References

[1] Tester, J. W, et al. The Future of Geothermal Energy. Massachusetts Institute of Technology; Cambridge, MA, USA. (2006).

[2] Clark, J. B. A hydraulic process for increasing the productivity of wells. Journal of Petroleum Technology (1949). , 1(1), 1-8.

[3] Green, C. A, Barree, R. D, & Miskimins, J. L. Hydraulic-Fracture-Model Sensitivity Analysis of a Massively Stacked, Lenticular, Tight Gas Reservoir. SPE Production & Operations (2009). , 24(1), 66-73.

[4] Behrmann, L. A, & Elbel, J. L. Effect of Perforations on Fracture Initiation. Journal of Petroleum Technology (1991). , 43(5), 608-615.

[5] De Pater, C. J, Cleary, M. P, Quinn, T. S, Barr, D. T, Johnson, D. E, & Weijers, L. Experimental Verification of Dimensional Analysis for Hydraulic Fracturing. SPE Production & Facilities (1994). , 9(4), 230-238.

[6] Ishida, T, Chen, Q, Mizuta, Y, & Roegiers, J. Influence of Fluid Viscosity on the Hydraulic Fracturing Mechanism. Transactions of the ASME; (2004). , 190-200.

[7] Wieland, C. W, Miskimins, J. L, Black, A. D, & Green, S. J. Results of a Laboratory Propellant Fracturing Test in a Colton Sandstone Block. Proceedings SPE Annual Technical Conference and Exhibition, September (2006). San Antonio, Texas, USA., 24-27.

[8] Macartney, H, & Morgan, P. The Potential for Geothermal Energy Recovery from Enhanced Geothermal Systems in the Raton Basin of Southern Colorado, USA. Proceedings AAPG Hedberg Geothermal Conference, Napa, CA, USA (2011).

[9] EMI Brazilian Tensile Strength Test Datasheet for Orica Core ID 4, Colorado School of Mines, Earth Mechanics Institute (2010).

[10] EMI Uniaxial Compressive Strength Test Datasheet for Orica Core ID 4, Colorado School of Mines Earth Mechanics Institute (2010).

[11] Gutierrez, M, Frash, L, & Hampton, J. Hydraulic Fracturing in Acrylic with Proppant. http://youtu.be/rbE4nisWlyAaccessed 10 October (2012).

[12] Gutierrez, M, Frash, L, & Hampton, J. Water Clear Acrylic Laboratory Hydraulic Fracturing Test. http://youtu.be/PEXOE2FTDlI.accessed 10 October (2012).

[13] ASTM ASTM D341: Standard Practice for Viscosity-Temperature Charts for Liquid Petroleum Products. Annual Book of ASTM Standards. ASTM International; West Conshohocken, PA. (2009).

[14] Ohtsu, M. Acoustic Emission Theory for Moment Tensor Analysis. Research in Nondestructive Evaluation (1995). , 6(3), 169-184.

[15] Weijers, L, De Pater, C. J, Owens, K. A, & Kogsbøll, H. H. Geometry of Hydraulic Fractures Induced From Horizontal Wellbores. SPE Production and Facilities (1994). , 9(2), 87-92.

[16] Swenson, D, Schroeder, R, Shinohara, N, & Okabe, T. Analysis of the Hijiori Long Term Circulation Test. Proceedings from the Twenty-Fourth Workshop on Geothermal Reservoir Engineering, January (1999). Stanford University, Stanford, CA, USA , 1999, 25-27.

Experimental Geomechanics

Chapter 8

Formation Fracturing by Pore Pressure Drop (Laboratory Study)

Sergey Turuntaev, Olga Melchaeva and
Evgeny Zenchenko

Additional information is available at the end of the chapter

Abstract

Pore pressure increase in saturated porous rocks may result in its fracturing and corresponding microseismic event occurrences. Another type of the porous medium fracturing is
related with rapid pore pressure drop at some boundary. If the porous saturated medium
has a boundary where it directly contacted with fluid under the high pore pressure, and the
pressure at that boundary is dropped, the conditions for tensile cracks can be achieved at
some distance from the boundary. In the paper, the results of experimental study of fracturing of the saturated porous artificial material due to pore pressure rapid drop are presented.
It was found that multiple microfracturing occurred during the pore pressure dropping,
which is governed by pore pressure gradient. Repeated pressure drops result in subsequent
increase of the sample permeability. The permeability was estimated on the basis of non-linear pore-elasticity equation with permeability dependent on pressure. The implementation
of calculations to laboratory experiment data showed significant variation of the porous
sample permeability during the initial non-stationary stage of the fluid pressure drop. The
acoustic emission activity variation was found to be controlled by pore pressure gradient
and changes of the number of potential fractures, which can be activated by the pore pressure gradient. It was found, that the probability distribution of these "potential fractures"
could be approximated by a Weibull distribution. A way of solution of the inverse problem
of local permeability defining from microseismic activity variation in a particular volume of
porous medium was suggested.

1. Introduction

Pore pressure change in saturated porous rocks may result in the rocks deformations and fracturing [1] and corresponding microseismic event occurrences. Microseismicity due to fluid injection is considered in numerous papers [2]. Another type of the porous medium fracturing is related with rapid pore pressure drop at some boundary. The mechanism of such fracturing was considered by [3] as a model of sudden coal blowing and by [4] as a model of volcano eruptions. If the porous saturated medium has a boundary where it directly contacted with fluid under the high pore pressure (in a hydraulic fracture or in a borehole), and the pressure at that boundary is dropped, the conditions for tensile cracks can be achieved at some distance from the boundary as it was shown by [3]. The effective stresses in the solid matrix will change with the speed of elastic waves, while the pore pressure changes will be governed by a kind of pore pressure diffusivity law. The phenomenon was studied by [4] in laboratory experiments with artificial material with high porosity filled by gas.

In the paper, the results of experimental study of fracturing of the porous sample saturated by fluid due to pore pressure rapid drop are presented. It was found that multiple microfracturing occurred during the pore pressure dropping, which is governed by pore pressure gradient variation. The locations of microcracks were found with the help of acoustic pulses recording. It was found that repeated pressure drops result in subsequent increase of the sample permeability. The permeability was estimated on the basis of non-linear pore-elasticity equation.

A mathematical model of the pore pressure variations was constructed based on pore pressure diffusion equation with diffusivity coefficient dependent on space and time. The implementation of analytical estimates and numerical calculations to laboratory experiment data showed significant variation of the porous sample permeability during the initial non-stationary stage of the fluid pressure drop. The acoustic emission activity variation can be described as a triggering process controlled by pore pressure gradient and changes of the number of potential fractures, which can be activated by the pore pressure gradient. It was found, that the probability distribution of these "potential fractures" could be approximated by a Weibull distribution. It was shown that it is possible to solve the inverse problem of defining local permeability from registered microseismic activity variation in a particular volume of porous medium.

2. Experimental procedure

The diagram and the photo of the experimental setup are shown in Figure 1. The samples were made of quartz sand with grain sizes 0.3...0.4 mm, the sand was cemented by "liquid glass" glue with mass fraction 1%. To prepare the sample, the sand/"liquid glass"/water mixture was tamped to a height 82 mm into a mould with 60 mm in inner diameter. Then it was dried during a week. The sample porosity was 35%, uniaxial unconfined compression strength was measured to be 2.5 MPa, p-wave velocity measured in the sample saturated by oil was 3100 m/s. The initial permeability measured by air blowing through the sample was about 2 D. To prevent the sample displacement during the experiments, a plastic ring was placed between lower end of the sample and the mould lid (see Figure 1). The upper end of the sample contacted directly with the mould upper lid. After vacuumization, the mould with sample was filled by mineral oil, which penetrates into the sample. Pressure in the mould was increased by means of oil injection through the bottom nipple up to 10 MPa and then discharged with the help of solenoid valve connected to the nipple. Injection-pressure drop cycles were repeated up to 40 times. Pressure release rate was controlled by a hydraulic resistor placed prior to the valve. The sample loading was related to oil pressurization at the bottom of the sample, there was no additional load.

Pore pressure transducers and acoustic emission (AE) transducers (of piezoceramic type) were mounted into upper and bottom lids of the mould, as it is shown in Figure 1. AE data were digitized with sampling frequency 2.5 MHz, the fluid pressure – with sampling frequency 50 kHz. The acoustic emission records were synchronized with the pressure records. Locations of AE events were estimated on the basis of measured p-wave velocity and onset time difference of AE pulses. The absolute peak amplitude of AE pulse was assumed as the amplitude of AE event. All localized AE events were characterized by onset time, location (distance from open boundary of the sample), and amplitude.

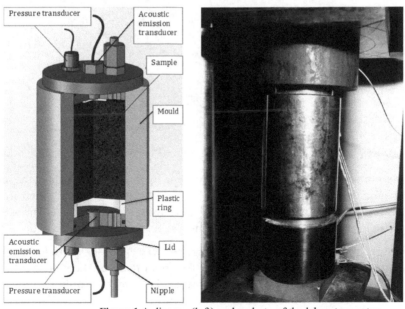

Figure 1 A diagram (left) and a photo of the laboratory setup.

Figure 1. A diagram (left) and a photo of the laboratory setup.

3. Results

Figure 2 shows typical waveforms of AE pulses, registered by opposite transducers. These waveforms had onsets with the same signs as well as with opposite signs. In case of tensile fractures it can be explained by the sample unloading in the processes of pore pressure drop and fracturing, so at least one or both boundaries of the tensile fracture were moving in the open end direction. After some time, the onsets of the AE pulses registered at the closed end became mainly positive. AE pulse amplitudes became lower with time.

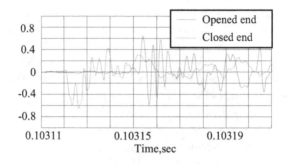

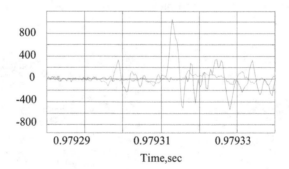

Figure 2. Examples of waveforms, registered at opposite sides of the sample. Amplitude is in arbitrary units.

Distributions of the AE pulse amplitudes summarized by 5 mm intervals along the sample are shown in Figure 3 for experiments after the 1st and 7th pressure drops. One can see, that microfracturing is spreading from the open end to the closed end with repeated pressure drops, and that maximal amplitudes registered at some distance from the opened end.

Variations of AE rate (the number of AE pulses per 0.1sec) are shown in Figure 4. The AE almost stops after 2 sec from the beginning of the pressure drop. The number of AE pulses increases with every next pressure drop, meanwhile the number of pulses with high amplitudes diminishes with next pressure drops. The last result can be explained by diminishing of pressure gradient due to the sample permeability increase in the course of subsequent fracturing of the sample.

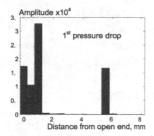

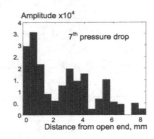

Figure 3. Distributions of the AE pulse amplitudes along the sample in two experiments.

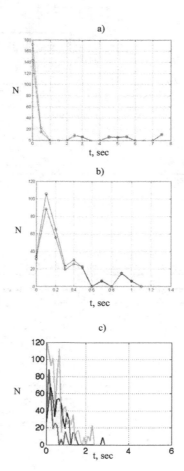

Figure 4. AE activity variations: a,b – registered at open (upper curve) and closed ends (lower curve) of the sample in experiments 1 and 6; c – registered in three different experiments (see Fig.5 legend)

Fluid pressure variations with time are show in Figure 5 for both open and close ends of the sample. Initially, the pore pressure decreases rapidly (in 0.1 sec), after that it slowly diminishes to atmospheric pressure. The AE rate is significant in first 2 sec, some of acoustic pulses occurred in up to 10 sec (Figure 4). The pressure gradient (estimated as pressure difference divided by the sample length) is shown in Figure 6. The prolongation of the pressure gradient maximal values is in agreement with AE maximal meanings.

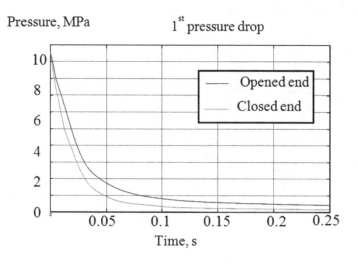

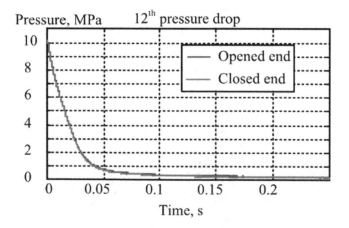

Figure 5. Fluid pressure vs. time registered at the two ends of the sample

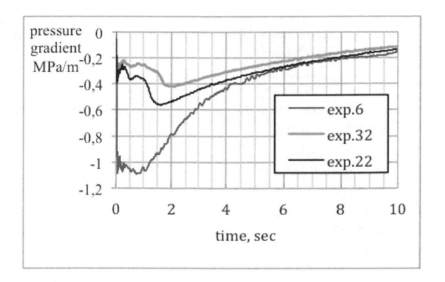

Figure 6. Linear estimation of the fluid pressure gradient for three experiments

4. Discussion

Shapiro et al. [5] proposed to consider an evolution of hydraulically induced microseismic event hypocenter locations as a diffusion process controlled by pore pressure diffusion in poro-elastic medium caused by fluid injection. In the presented experiments the fluid pressure decreased with time and AE maximum was registered when the pressure was dropped from its maximum. Let us compare AE variation and variation of pressure difference (Figure 4 and Figure 6). One can see that the AE began when the absolute value of the fluid pressure difference started to increase, and AE stops when the fluid pressure difference started to decrease. It is clear from physical point of view, that to produce tensile fracturing one should have tensile force, which could appear only in case of enough high pressure gradient.

Let us now try to estimate dynamic permeability variation during the fluid pressure drop. For it, we used fluid pressure data registered at the open end of the sample and calculated the fluid pressure at the closed end of the sample by means of simple pore-elastic equation (Schelkachev, [6]) for small time intervals (0.01 sec) and one-dimensional isotropic homogeneous case:

$$\frac{\partial p}{\partial t} = D\frac{\partial^2 p}{\partial x^2}$$

where D is hydraulic diffusivity

$$D = \frac{k}{\mu_0 \beta m_0}$$

where k is permeability, β is an effective compressibility of the porous medium, μ_0 – viscosity, m_0 – initial porosity. Initial condition:

$$p(x, 0) = p_0$$

A zero fluid rate Q at the closed end of the sample and registered pressure at the open end were taken as boundary conditions:

$$\frac{\partial p}{\partial x}(t, l) = 0$$

$$p(t, 0) = p_1(t)$$

Then we compared results of the calculations with experimental data and vary coefficient of diffusivity to obtain the best coincidence between calculated and registered pressure. The procedure was repeated for all the time of experiment. There was no difference between experimental and calculated pressures, obtained by that manner (Figure 7). The dependence of estimated permeability on fluid pressure is shown in Figure 8. The final permeability values after the pressure drop are shown in Table 1.

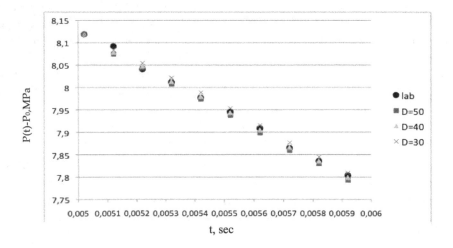

Figure 7. Measured (lab) and calculated for several diffusivity coefficients differential pore pressure.

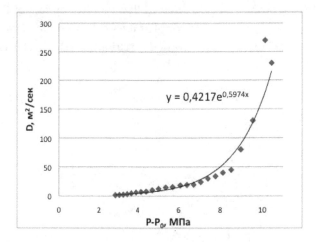

Figure 8. Estimation of the permeability dependence on pore fluid pressure

Pressure drop #	6	22	32
Permeability k, D	4.9	7.4	9.5

Table 1. Permeability estimated by pore pressure difference diminishing in time.

To find AE relation with the pore pressure gradient the following assumption can be used [5, 7]:

- AE event occurred when the pore pressure gradient reaches some critical value;

- The critical value varies spatially and can be described by a probability distribution.

One can suggests that the critical value distribution can be described by Weibull distribution which is often used to describe fragment size distributions in fractured rock [8];

$$N((dp/dx)^*) = N * ba^{-b}((-dp/dx)^*)^{b-1}e^{\left(\frac{(dp/dx)^*}{a}\right)^b},$$ (1)

where parameters a and b are the scale and the shape parameters, respectively, $(dp/dx)^*$ is the critical value of pore pressure gradient.

Variation of AE rate in time can be described with the help of Weibull distribution

$$N(t) = N * \frac{c}{t}\left(\frac{t}{t_0}\right)^c e^{-\left(\frac{t}{t_0}\right)^c}$$ (2)

The parameters of the distribution calculated to fit experimental data of the experiment 22 (Figure 9) are $N^*=120$, $c=1,7$, $t_0=0,6$. The coupled use of distribution (2) and pore pressure measurement allows to calculate parameters of the distribution (1) to best fit experimental data of $(dp/dx)^*(t)$.

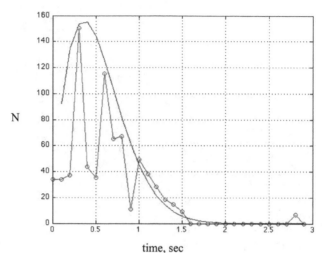

Figure 9. Variation of AE rate and fitted Weibull function (experiment 22).

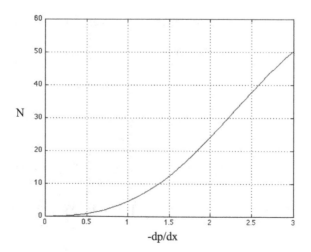

Figure 10. Dependence of AE event numbers on pore pressure gradient calculated in accordance with relation (1) with parameters of best fit curve shown in Figure 9.

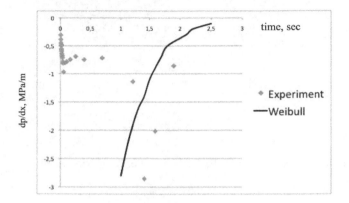

Figure 11. Experimental and calculated pore pressure gradient variation in time (experiment 22).

The comparison of experimental pore pressure gradient (estimated as pressure difference between two points of pressure measurements) and critical pore pressure gradient calculated based on AE variations is shown in Figure 11. Experimental and calculated data start to be in agreement when the fluid pressure become respectively low and the sample permeability is more or less constant.

So, if one gets to know variation of AE (or microseismic activity in real case) in time and the relation between number of events and critical values of pore pressure gradient is known, it is possible to calculate the porous medium permeability.

Let's now consider once more one-dimensional pore-elasticity equation with constant coefficient of diffusivity

$$\frac{\partial p}{\partial x} = D\frac{\partial^2 p}{\partial x^2} \tag{3}$$

The solution can be written as

where
$$p(x,t) = A + \sum_{i=0}^{\infty} C_i e^{-\mu_i^2 Dt} \sin(\mu_i x) + \sum_{i=0}^{\infty} B_i e^{-\mu_i^2 Dt} \cos(\mu_i x) \tag{4}$$

Use of initial and boundary conditions of the experiment (which were described early) the solution can be written as

$$p(x,t) = p_{atm} + A(x)\sum_{i=0}^{\infty} \frac{1}{\mu_i^2} e^{-\mu_i^2 Dt} \cos(\mu_i x) \tag{5}$$

where

$$A(x) = \frac{p(x, 0) - p_{atm}}{\sum\limits_{i=0}^{\infty} \frac{1}{\mu_i^2} \cos(\mu_i x)}, \quad \mu_i = \pi i + \frac{\pi}{2}$$

The series

$$\sum\limits_{i=0}^{\infty} \frac{1}{\mu_i^2} \cos(\mu_i x)$$

converges and majorizes the series in (5). Let's consider a function $f(x)$:

$$f(x + T) = f(x), \quad T = 4;$$
$$f(-x) = f(x);$$
$$\forall x \in [0;2]: f(x) = \frac{1-x}{2}.$$

For $f(x)$ defined at $[-l,l]$ with period $2l$ which satisfies the Dirichlet conditions, the Fourier series expansion looks like

$$f(x) = \frac{a_0}{2} + \sum\limits_{n=1}^{\infty} (a_n \cos\frac{\pi n x}{l} + b_n \sin\frac{\pi n x}{l}),$$

$$a_n = \frac{1}{l} \int\limits_{-l}^{l} f(x) \cos\frac{\pi n x}{l} dx$$

$$b_n = \frac{1}{l} \int\limits_{-l}^{l} f(x) \sin\frac{\pi n x}{l} dx$$

The function $f(x)$ is even, so

$$b_n = 0$$

$$a_0 = \frac{1}{2} 2 \int\limits_0^2 \frac{1-x}{2} dx = 0$$

$$a_n = \frac{1}{2} 2 \int\limits_0^2 \frac{1-x}{2} \cos(\frac{\pi n x}{2}) dx = \frac{1}{2}([(1-x)\frac{2}{\pi n}\sin(\frac{\pi n x}{2})]_0^2 - \int\limits_0^2 (-1)\frac{2}{\pi n}\sin(\frac{\pi n x}{2}) dx) =$$

$$= \frac{1}{\pi n} \int\limits_0^2 \sin(\frac{\pi n x}{2}) dx = \frac{2}{\pi^2 n^2}(-1)\cos(\frac{\pi n x}{2})|_0^2 = \frac{2}{\pi^2 n^2}(1 - (_1^-))$$

$$f(x) = \sum\limits_{n=1}^{\infty} \frac{2}{\pi^2 n^2}(1 - (-1)^n)\cos(\frac{\pi n x}{2})$$

Substitution $n = 2k + 1$ (for even n $f(x)=0$)

$$f(x) = \sum_{k=0}^{\infty} \frac{4}{\pi^2 (2k+1)^2} \cos\left(\frac{\pi(2k+1)x}{2}\right) = \sum_{k=0}^{\infty} \frac{1}{(\pi k + \frac{\pi}{2})^2} \cos\left((\pi k + \frac{\pi}{2})x\right)$$

$$A(x) = \frac{p(x,0) - p_{atm}}{\sum_{i=0}^{\infty} \frac{1}{\mu_i^2} \cos(\mu_i x)}, \quad \mu_i = \pi i + \frac{\pi}{2}$$

so:

$$A(x) = \frac{p(x,0) - p_{atm}}{\frac{1}{2}(1-x)} \tag{6}$$

To estimate the permeability one can adopt that

$$\frac{\partial A}{\partial x}(x) = 0$$

because in considered experiments

$$\frac{\partial p}{\partial x}(x,0) = 0$$

and denominator in (6) is almost constant when x is small. In that case:

$$\frac{\partial p}{\partial x} = -A(x) \sum_{i=0}^{\infty} \frac{1}{\mu_i} e^{-\mu_i^2 Dt} \sin(\mu_i x) \tag{7}$$

The diminishing part of the pore pressure gradient dependence on time (which is shown by ellipse in Figure 12) can be approximated by exponential function be^{at} (as it is shown in Figure 13), which corresponds to $i=0$ in (7), and the diffusivity coefficient can be estimated as

$$D = \frac{a}{(\pi/2)^2}$$

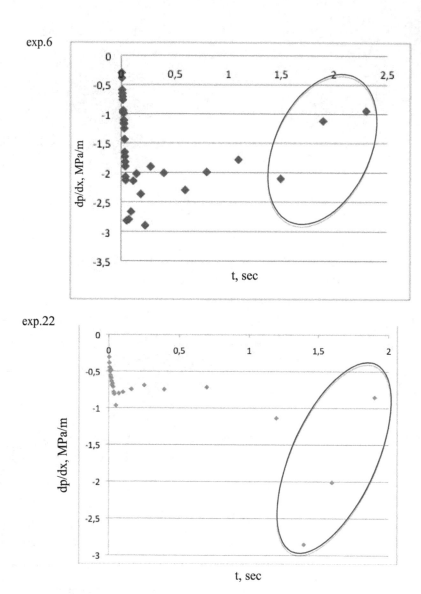

Figure 12. Variation of pressure difference in time. The ellipse shows data used for permeability calculation.

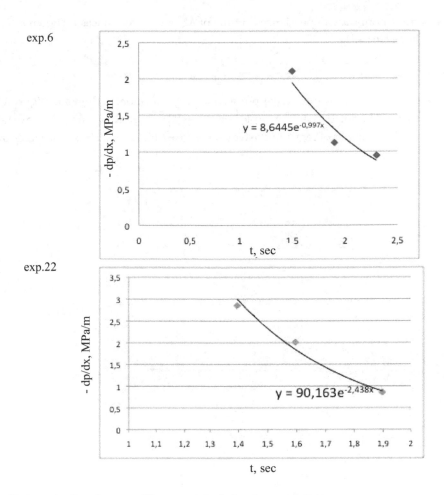

Figure 13. Approximation of pressure difference variation in time by exponential low.

The results of the permeability estimations are shown in Table 2.

Pressure drop #	6	22	32
D, m²/sec	0.4	0.99	0.785
k, D	3.8	9.2	7.3

Table 2. Estimated values of diffusivity coefficient and permeability

The obtained values of permeability were compared with permeability estimated by permeability dependence on pore pressure drop (Table 1) and with permeability obtained with the help of r-t method suggested by Serge Shapiro and colleagues ([5, 9-11]). Diagram of distance from the sample boundary dependence on time of AE event occurrences is shown in figure 14. An envelope curve.

$$r = \sqrt{4\pi D(t - t_s)},$$

which is a solution of one-dimensional porous elasticity equation, is shown in Figure 14 by red curve. It should be noted that the sample length was 83 mm, which is shown in the diagram by dotted line. The number of registered AE events was not high, it restricts an accuracy of the method.

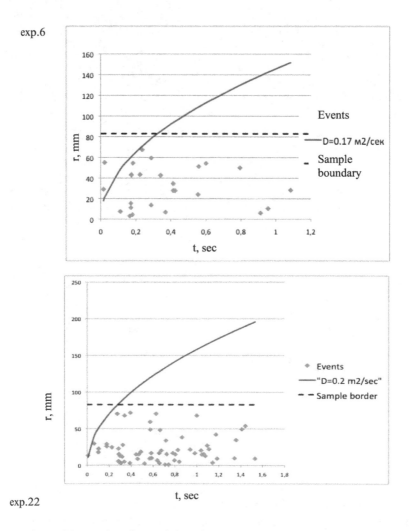

Figure 14. Dependence of distance from the sample boundary on time of AE event occurrences.

If we compare Table 1, Table 2 and Table 3 we will see, that the permeability based on AE variation in time estimation gives values which are close to that obtained using data on pore pressure drop; nevertheless, r-t method gives values which are not contradict other results but differ from them significantly.

Experiment #	2	6	22	32
D, m²/sec	0.003	0.17	0.2	2
k, D	0.03	1.6	1.9	18.6

Table 3. Permeability estimated by r-t method

5. Conclusions

In the paper, the results of experimental study of saturated porous sample fracturing due to pore pressure rapid drop are discussed. It was found, that acoustic emission AE (corresponded to microfracturing) was spreading from the end of the pressure drop to the closed end of the sample, and that maximal number of AE events was registered at some distance from the opened end.

The number of AE pulses increased with every next pressure drop, meanwhile the number of pulses with high amplitudes diminished. The prolongation of the pressure gradient maximal values is in agreement with AE maximal rate.

It was found that multiple microfracturing occurred during the pore pressure drop; the microfractiring is governed by pore pressure gradient.

The model of AE relation with the pore pressure gradient was considered based on the following assumptions: AE event occurred when the pore pressure gradient reaches some critical value; the critical value varies and can be described by Weibull distribution, which is often used to describe fragment size distributions in fractured rocks.

Permeability variation during the fluid pressure drop was estimated by means of fluid pressure data and pore-elastic equation solution for small time intervals (0.01 sec). It was found that the sample permeability is high in initial stage of the pressure discharge and decrease during pore pressure drop.

It is shown that if the change in microseismic activity in time is measured, the distribution of the critical pressure gradient is known for the considered material and the boundary conditions are given (for example, the change in pressure in the well), it is possible to calculate the pressure gradient, and on this basis, the permeability of the porous medium.

The study showed possibility to solve an inverse problem of defining permeability by registering microseismic activity variation in particular volume of porous medium alongside with pore pressure measurements at some point.

Acknowledgements

The idea of the study was suggested by Dimitry Chuprakov. The authors wish to acknowledge the generous support of Russian Foundation for Basic Research (RFBR project # 10-05-00638) and of the Russian Academy of Sciences Presidium Program #4.

Author details

Sergey Turuntaev[1*], Olga Melchaeva[2] and Evgeny Zenchenko[1]

*Address all correspondence to: stur@idg.chph.ras.ru

1 Institute of Geosphere Dynamics of Russian Academy of Sciences (IDG RAS), Moscow, Russia

2 Moscow Institute of Physics and Technology, Moscow, Russia

References

[1] Maury, V. et D. Fourmaintraux. Mecanique des roches appliquee aux problemes d'exploration et de production petrolieres. Societe Nationale Elf Aquitaine (Production). Boussens. (1993).

[2] Maxwell, S. Microseismic: Growth born from success. The Leading Edge (2010). , 3-338.

[3] Khristianovich, S. A. Transient liquid and gas flow in porous medium under abrupt pore change in time and high rate gradient. Journal of Mining Science (1985). , 1-18.

[4] Alidibirov, M, & Panov, V. Magma fragmentation dynamics: experiments with analogue porous low- strength material. Bull Volcanol. (1998). , 59-481.

[5] Shapiro, S. A, Rothert, E, Rath, V, & Rindschwentner, J. Characterization of fluid transport properties of reservoirs using induced microseismicity: Geophysics, (2002). , 67-212.

[6] Schelkatchev, V. N. Fundamentals and applications of nonstationary filtration theory: Moscow, Neft i gaz, (1995).

[7] Turuntaev, S. B, Eremeeva, E. I, & Zenchenko, E. V. Laboratory study of microseismicity spreading due to pore pressure change. Journal of Seismology (2012). DOIs10950-012-9303-x.

[8] Tsvetkov, V. M, Lukishov, B. G, & Livshits, L. D. Fragment formation in crushing a brittle medium under hydrostatic compression. Journal of mining science, (1979). DOI:10.1007/BF02539986., 15(3), 228-232.

[9] Shapiro, S. A, Dinske, C, & Rothert, E. Hydraulic-fracturing controlled dynamics of microseismic clouds. Geophysical Research letters (2006). L14312.

[10] Dinske, C, Shapiro, S. A, & Rutledge, J. T. Interpretation of microseismicity resulting from gel and water fracturing of tight gas reservoirs. Pure Applied Geophysics (2009). DOI:s00024-009-0003-6.

[11] Grechka, V, Mazumdar, P, & Shapiro, S. A. Predicting permeability and gas production of hydraulically fractured tight sands from microseismic data. Geophysics (2010). BB10, doi:, 1.

Chapter 9

Comparison of Hydraulic and Conventional Tensile Strength Tests

Michael Molenda, Ferdinand Stöckhert,
Sebastian Brenne and Michael Alber

Additional information is available at the end of the chapter

Abstract

Tensile strength is paramount for reliable simulation of hydraulic fracturing experiments on all scales. Tensile strength values depend strongly on the test method. Three different laboratory tests for tensile strength of rocks are compared. Test methods employed are the Brazilian disc test (BDT), modified tension test (MTT) and hydraulic fracturing experiments with hollow cylinders (MF = Mini Frac). Lithologies tested are a micritic limestone, a coarse-grained marble, a fine-grained Ruhrsandstone, a medium-grained rhyolite, a medium- /coarse-grained andesite and a medium grained sandstone. Test results reveal a relationship between the area under tensile stress at failure and the measured tensile strength. This relationship becomes visible when the area under tensile strength ranges over one order of magnitude from 450 to 4624 mm². This observation becomes relevant when selecting the tensile strength values of lithologies.

Keywords: hydraulic fracturing, Brazilian Disc test, Modified Tension Test, Acoustic Emission, numerical simulation

1. Introduction

Tensile strength tests are widely applied in rock mechanics to obtain input parameters for planning of hydraulic fracturing on all scales. In literature only few experimental data sets are published dealing with samples size effects on tensile strength tests [1,2] or the comparison of different tensile tests in general [1,3]. Usually, results of laboratory tensile tests are taken to be size independent when used as input parameter for numerical studies at different spatial sizes.

We compare the results of 3 different, easily applicable laboratory tests for tensile strength of rocks. The sample set comprises a micritic limestone, a coarse-grained marble, a fine-grained Ruhr-Sandstone, a medium-grained rhyolite, a medium- /coarse-grained andesite and a medium grained sandstone. All tested rocks were characterized petrographically as well as by ultrasonic velocities, density, porosity, permeability, static, dynamic elastic moduli and compressive strength. In order to determine the effects of specimen size on test results, we carried out BDT according to ISRM [4] with disc diameters of 30, 40, 50, 62, 75 and 84 mm, respectively. The recently presented MTT [5] was used as a tensile strength test with an approximately uniform tensile stress distribution. Hydraulic tensile strength was evaluated by MF experiments (core diameter 40 and 62 mm; borehole/diameter ratio 1:10) under uniaxial compression [6]. MF pressurization was performed with a constant fluid volume rate of 0.1 ml/s representing a stress rate of 0.3 MPa/s. In all tests relevant acoustic emission (AE) values have been evaluated to get additional information on the failure processes.

2. Materials and methods

2.1. Sample material

To investigate the influence of rock properties on tensile test methods, six different rock types were tested. Bebertal sandstone, a medium grained Permian sandstone from a quarry near Magdeburg, Germany. Ruhrsandstone, a fine-grained and massive Carboniferous arcose from the Ruhr area in Germany. A medium to coarse grained, jointed Permian andesite from the Doenstedt Eiche quarry near Doenstedt, Germany. A medium grained, highly jointed Permian rhyolite from the Holzmuehlental quarry near Flechtingen, Germany. A micritic Jurassic limestone from a quarry near Treuchtlingen, Germany and a coarse grained marble from Carrara, Italy. The rocks' petrophysical properties, namely bulk density, grain density, compressional wave speed, porosity, permeability, cohesion and friction angle are listed in Table 1.

2.2. Petrophysical characterization

Dry densities are calculated geometrically based on geometrical properties, grain densities are measured according to DIN 18124. Compressional wave velocities are measured at each core with a Geotron USG 40/UST 50-12 at room temperature and in dry condition. Porosities are derived from the difference between grain density and geometrical density of the oven-dried samples. Permeabilities are evaluated via a constant head test on the hollow cylinder samples used for the MF tests [7]. Bebertal-sandstones are permeable enough to use a simple axial flow-through test with a maximum pressure difference of up to 3 bars. The samples are sealed off with rubber jackets to minimize water-flow along the sample surface. Unconfined compressive strengths and static moduli of elasticity are measured by uniaxial compressive tests [8].

Rock type (location)	ρ_d [g/cm³]	ρ_s [g/cm³]	vp [m/s]	Φ [%]	k [m²]	c [MPa]	φ [°]
Marble (Carrara)	2.71 ±0.002	2.721 ±0.003	5.67 ±0.06	0.40	1E-19	29	22
Limestone (Treuchtlingen)	2.56 ±0.008	2.713 ±0.002	5.59 ±0.05	5.64	1E-18	27	53
Ruhrsandstone (Ruhr area)	2.57 ±0.006	2.688 ±0.008	4.61 ±0.13	4.39	8E-18	36	50
Rhyolite (Flechtingen)	2.63 ±0.015	2.657 ±0.011	5.39 ±0.34	1.02	9E-19	20-36	55
Andesite (Dönstedt)	2.72 ±0.023	2.734 ±0.006	5.26 ±0.28	0.51	-	20-41	50
Sandstone (Bebertal)	2.66 ±0.061	2.44 ±0.059	3.61 ±0.61	8.27	11E-15	15	45

Table 1. Averaged values of petrophysical properties of the rock samples. ρ_d dry bulk density, ρ_s grain density, vp compressional wave velocity, Φ porosity, k permeability, c cohesion, φ friction angle.

2.3. Testing procedure of the tensile strength tests

All experiments are performed in a stiff servo-hydraulic loading frame from Material Testing Systems (MTS) with a load capacity of 4000 kN. For further details on the technical specifications see Table 2.

Device (manufacturer) name	max. capacity	accuracy	BDT	MTT	MF
Axial load cell (Althen) CPA-50	500 KN	± 100 N	x	x	x
Axial displ. transducer (Scheavitz) MHR 250 LVDT 1 & 2	6.3 mm	$\pm 1 \bullet 10^{-4}$ mm	x	x	x
Displ. transducer at pressure intensifier (HBM) WA 100 mm LVDT 3	100 mm	$\pm 1 \bullet 10^{-3}$ mm			x
Load cell for Hoek Cell					
Load cell for pressure intensifier (Burster) 8219R-3000	300 MPa	± 0,03 MPa			x

Table 2. Technical specifications of the measurement system.

Acoustic Emission (AE) signals are acquired with an AMSY-5 Acoustic Emission Measurement System (Vallen Systeme GmbH, Germany) equipped with up to 6 Sensors of type VS150-M. The Sensors are sensitive in a frequency range of 100-450 kHz with a resonance frequency of 150 kHz and a preamplification of 34 dB$_{AE}$. Due to machine noise in the range below 100 kHz incoming signals are filtered by a digital bandpass-filter in a frequency range of 95-850 kHz.

AE data are sampled by a sampling rate of 10 kHz. The sensors are fixed using hot-melt adhesive to ensure best coupling characteristics. Pencil-break tests (Hsu-Nielsen source) and sensor pulsing runs (active acoustic emission by one sensor) are used to ensure good sensor coupling of the sensor on the sample.

2.3.1. Hydraulic fracturing core experiments (MF) procedure

Minifrac experiments are carried out mainly on 40 mm cores with a borehole diameter of 4mm. Furthermore some 62 mm cores with a borehole of 6 mm diameter are tested. The samples are loaded axially up to 5 MPa to ensure that the packer mechanism is tight and seals off the borehole openings at the top and at the bottom. The borehole pressure was raised servo controlled with a fixed volume rate of 0.1 ml/s that results in a pressure rate of approximately 0.3 MPa/s. All MF tests are monitored by Acoustic Emissions with four sensors glued directly to the samples and a fifth sensor placed at the incoming hydraulic line.

2.3.2. Brazilian Disc Tests (BDT) procedure

All Brazilian disc tests are carried out following the ISRM suggested method [4] at a load rate of 200 N/s. Disc diameters used are 30, 40, 50, 62, 75 and 84 mm, whereas the length to diameter ratio (L/D) was constant at 0.5. All tests are monitored by one AE-sensor glued directly in the middle of the disc specimen. The size dependency is tested with discs from Ruhrsandstone, marble, rhyolite and limestone.

2.3.3. Modified Tension Test (MTT) procedure

The MTT tests are driven load controlled at a rate of 200 N/s that corresponds to a stress rate of 0.02 MPa/s. The axial force is applied from the top (Figure 1). MTT test samples are observed by up to 6 AE-Sensors glued directly to the specimen. The samples were overcored with 62 mm and 30 mm diameters where the overlapping height is 1/3 of the total sample height (Figure 1). The centralizing of the drills was achieved by using a former plate to adjust the sample before drilling. Despite assiduously arrangement the eccentricity of the overcoring was in the range of up to 3 mm due to the imprecise vertical guidance of a standard drilling machine. In order to test the influence of eccentricity we also prepared samples with an eccentricity of 14 mm.

3. Experimental results

3.1. Brazilian disc test size dependency

The size dependency of the absolute size of the Brazilian disc test discs on the tensile strength is shown in Figure 2. Overall 138 Brazilian disc tests are undertaken for up to 6 sizes and four lithologies. The disc diameters, ranging from 30-84 mm, represent the sizes that are mostly tested in laboratories to determine the BDT tensile strength of rock samples. The results of the size dependency tests show no significant relationship between the sizes of the tested disc to

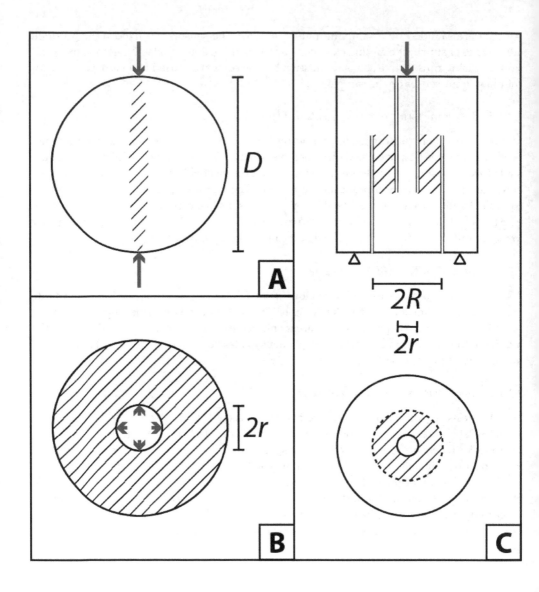

Figure 1. Sketches of the three tensile test methods. A: BDT side view, B: MF top view, C: MTT side view cross section (upper) and top view (lower).

its calculated tensile strength as long as the length to diameter ratio is held constant as suggested by the ISRM suggested method at a value of 0.5 [4]. There is a marginal tendency for the standard deviation of the tensile strength to decrease with increasing disc size.

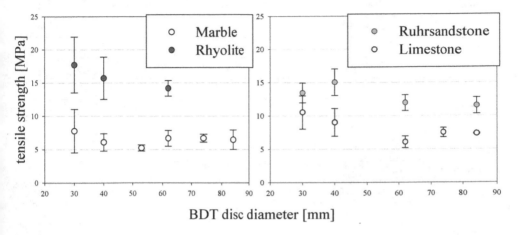

Figure 2. Size dependency of the BDT disc size on the tensile strength for four lithologies. Circles represent the mean values, bars stand for the standard deviation.

3.2. MF, BDT and MTT tensile strength results

Three different methods for the determination of tensile strength are compared regarding their results. 201 Brazilian disc tests, 31 Minifrac tests and 15 Modified tension tests form the basis of the data evaluation, where σ_t^{BDT}, σ_t^{MF} and σ_t^{MTT} are the tensile strengths indexed by the used method. BDT tensile strengths are calculated as follows [4].

$$\text{BDT: } \sigma_t^{BDT} = 2P / \pi Dt \tag{1}$$

Where P is the force at failure, D is the disc diameter and t the disc thickness.

For the MF tests, assuming the rocks to be nearly impermeable and therefore neglecting a relevant pore pressure influence the tensile strength is given directly by the breakdown pressure P_b [9].

$$\text{MF: } \sigma_t^{MF} = P_b \tag{2}$$

MTT tensile strengths are evaluated by the formula given by [5].

$$\text{MTT: } \sigma_t^{MTT} = F_{max} / A_{TZ} = F_{max} / (R^2 \pi - r^2 \pi) \tag{3}$$

Where R^2 and r^2 are the outer and inner radius, respectively (Figure 1). Mean values, standard deviations and total number of tests for all three testmethods can be found in Table 3.

Lithology	Test method	Mean [MPa]	Std. dev. [MPa]	N [-]
	BDT	13.2	2.1	32
Ruhrsandstone	MF	19.0	3.0	10
	MTT	5.8	1.0	3
	BDT	15.8	3.2	39
rhyolite	MF	20.1	5.5	5
	MTT	4.9	1.4	2
	BDT	8.2	2.2	36
limestone	MF	10.2	1.7	5
	MTT	4.8	1.0	2
	BDT	6.4	1.5	32
marble	MF	7.8	1.3	4
	MTT	4.3	1.2	2
	BDT	14.6	4.5	23
andesite	MF	14.4	5.1	4
	MTT	8.7	4.4	3
	BDT	4.1	1.2	39
Bebertal sandstone	MF	4.3	2.0	3
	MTT	2.4	-	1
	MTT eccentric	1.0	5E-3	2

Table 3. Comparison of tensile strength out of three test methods.

One of the main observations is the very low tensile strength measured with the Modified Tension Test method. The MTT results mean values are in the range of 66 % down to 31 % of those obtained with the BDT. In addition to the low tensile strengths obtained by the MTT an eccentricity of the overcoring yields to an additional underestimation of the tensile strength values. The BDT and MF results seem to be more similar. The BDT results lie in the range of 70 % to 100 % of the MF tensile strength, so the MF test yields the highest tensile strengths and also to the highest standard deviations. All measurements are visualized in Figure 3. Doenstedt andesite and Flechtingen rhyolite tensile strengths have the highest standard deviations of the tested rock types. This variance is due to the high amount of natural joints that are assumed to have a different tensile strength with respect to the intact parts. Therefore the tensile strength scattering is the result of the material heterogeneity itself.

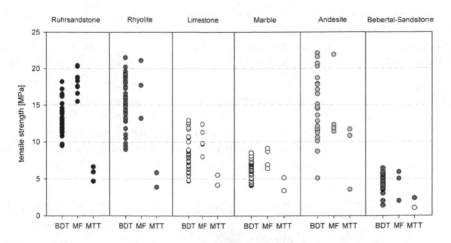

Figure 3. Results of all tensile strength test. BDT: Brazilian Disc Test, MF: Minifrac, MTT: Modified Tension Test. Hollow circle in Bebertal sandstone MTT tests represents two results of the highly eccentrical MTT tests.

3.3. Acoustic Emissions results

Acoustic emission data obtained during the tests give rough insights into the failure processes. It is obvious that all tests end with a spalling of the specimens in parts due to a complete tensile failure. Simple AE count analysis show that the BDT is accompanied with an immense hit-rate long before total failure in comparison to the relatively quiet pre-failure phases of the MF and MTT tests. In good agreement with theoretical considerations of the stress distribution in the Brazil disc [1] these events are most likely due to compressional failure at the top and bottom of the disc, accompanied with crack propagation and coalescence before peak load (Figure 4).

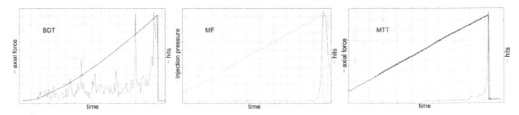

Figure 4. AE hits per 0.5 sec., BDT left, MF middle and MTT right showing the huge difference in AE hits before total failure of the sample.

4. Numerical model

We investigate the effect of eccentricity of the overcoring for the MTT samples by a numerical simulation. A finite element study that has been performed by Plinninger et al. [10] that shows a uniform tensile stress distribution in the annulus of the test samples. It is arguably if this model is the right tool for modeling a tensile stress distribution in rock samples prior to failure. A simple linear elastic 3D FEM model reveals tensile stress concentrations at the edges of the rims in the sample (Figure 5). Fractures may be initiate there at relative low axial forces.

During preparation of the samples it becomes obvious that exact centralization of the inner overcoring is not always given. Two Bebertal sandstone samples were prepared with a eccentricity of 14 mm resulting in a minimum rim width of 2 mm instead of 16 mm for a perfectly centralized sample. The average eccentricity of our samples is in the range of up to 3 mm. Tensile stress redistribution due to eccentricity is modeled as well and can easily double the tensile stress in the thinner rim of the annulus (Figure 5).

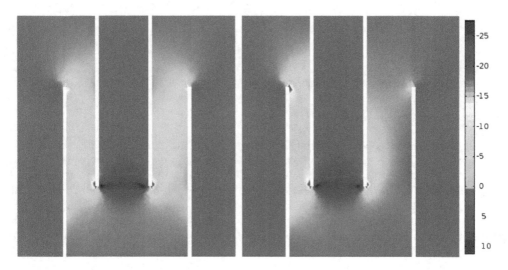

Figure 5. Slice through a linear elastic 3D FEM model of MTT tensile test. Values of axial stress are given in MPa where negative values stand for tensile stress. Left model represents a perfectly centralized sample. Right model shows the stress distribution for a eccentricity of 6 mm towards the left edge.

5. Discussion

247 tensile strength test results of BDT, MF and MTT tests vary considerable within one lithology (Figure 6). Therefore it is not trivial to give a reliable prediction of the tensile strength parameter. Results of the BDT tests show no significant variation with respect to the specimen size, as long as the aspect ratio is held constant. Nevertheless the tensile strength data scattering is high, so that it may obscure existing trends. Acoustic Emission evaluation shows that during the BDT multiple fracturing mechanisms are present. Before total fracturing of the sample by a tensile rupture there is a high amount of AE activity that is most likely related to compressional failure at the top and bottom of the disc. Beside this, compressional stress concentrations and the inhomogeneous tensile stress distribution may lead to tensile cracks before peak load.

MF results lead to the highest tensile strengths in this comparison where there seem to be no differences in tensile strength when using a 4 mm or a 6 m borehole for pressurization. Again one has to take into account that the high amount of tensile strength scattering for these tests inhibits a statement regarding a borehole size dependency.

The results of the MTT tests give the lowest tensile strengths and very low standard deviations. Latter may be related to the small amount of testes MTT per lithology. Furthermore all MTT are prepared using the same sample sizes. A major problem of the MTT experiments is the centralization of the boreholes. An eccentricity yields to a significant inhomogeneity of the tensile stress distribution in the sample (Figure 5). Numerical simulations of the MTT eccentricity effect together with the two eccentric MTT samples (Figure 3) show that the calculated tensile strength may be underestimated massively. One reason for the apparently lower tensile strength measured using the MMT might be the applicability of Equation (3). In deriving the equation, it was assumed that, when the peak load is approached, the tensile stress distribution is almost uniform in the area defined as A_{TZ} [5]. This may only be true if the material is highly ductile. However, for brittle rocks, especially for highly fractured rocks, fracture propagation may occur and lead to ultimate failure at a much lower load as suggested by Equation (3) due to stress concentration (Figure 5).

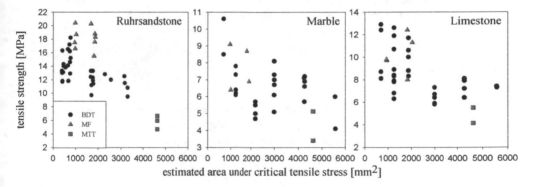

Figure 6. Tensile strength results plotted against the assumed area under tension. BDT: diameter x thickness, MF: surface area of the borehole and MTT: twice the surface area between the outer and inner borehole, upper and lower.

Testmethod	BDT	MF	MTT
Area under 1,51,5	450-3400 mm²	1005-3393 mm²	4624 mm²
Calculation	$D \cdot t$	$2 \cdot \pi \cdot rbh \cdot l$	$2 \cdot (\pi \cdot (R^2 - r^2))$

Table 4. Estimated area subjected to tensile stress for the different tensile tests. *D*: BDT disc diameter, *t*: BDT disc thickness, *rbh* : MF borehole radius, *l*: MF sample height, *R*: MTT outer borehole radius, *r*: MTT inner borehole radius.

Main difference in all experiments and the reason for choosing these are the areas that are under tensile stress at the point of failure. The calculated tensile strengths compared to the area perpendicular to the maximum tensile stress show a negative trend for the tensile strength with increasing area being set under tensile stress. That is reasonable in terms of the statistical theory of strength. Especially for the igneous rocks it seems evident, that the probability to set a healed joint under a critical tension rises with the size of the sample volume that is under tensional stress. For the selection of the tensile strength test one should keep in mind that depending on the lithology the apparent tensile strength appears to be a function of the area, or more exact of the volume under tensile stress. Thus, for a relative homogeneous rock a less severe reduction of the measured tensile strength with size will be visible as it will be at the highly fractured igneous rocks tested in this study.

It is arguable and may not be appropriate to study the effect of area/volume under tensile stress on the measured tensile strength using the combined results from different types of tests, especially if the different tests tend to give different average measured tensile strengths. Furthermore the negative trend of tensile strength with respect to the stressed area/volume is not that obvious for the single test methods. Especially the assumption of uniform tensile stress distribution close to peak load in the annulus [5] for the MTT samples seems not to be comprehensible. It may hold for ductile materials but not for brittle ones. Therefore the validity of equation (3) for the calculation of the tensile strength is questionable. Nevertheless the resulting tensile strengths are treated as the same rock property when used as input parameters for calculations. This is very problematic due to its huge variation as shown in the tests. The correlation of the calculated tensile strength with the stressed area/volume is one possible approach to account for the decreasing apparent tensile strength behavior.

Acknowledgements

This work is funded by the Federal Ministry of Environment, Nature Conservation and Nuclear Safety (funding mark 0325279B). Special thank goes to Kirsten Bartmann and Sabrina Hoenig for laboratory work and data evaluation done during their Master Theses.

Author details

Michael Molenda*, Ferdinand Stöckhert, Sebastian Brenne and Michael Alber

*Address all correspondence to: michael.molenda@rub.de

Ruhr-University Bochum, Germany

References

[1] Mellor, M, & Hawkes, I. Measurement of tensile strength by diametral compression of discs and annuli. Engineering Geology. Elsevier; (1971). , 5(3), 173-225.

[2] Thuro, K, Plinninger, R. J, Zäh, S, & Schütz, S. Scale effects in rock strength properties. Part 1: Unconfined compressive test and Brazilian test. Rock Mechanics-A Challenge for Society.-881 p., Proceedings of the ISRM Regional Symposium Eurock. (2001). , 169-74.

[3] Hudson, J. A, Brown, E, & Rummel, F. The controlled failure of rock discs and rings loaded in diametral compression. International Journal of Rock Mechanics and Mining Sciences & Geomechanics Abstracts. Elsevier; (1972). IN1-IN4, 245-8., 9(2), 241-4.

[4] Bieniawski, Z. T, & Hawkes, I. Suggested Methods for Determining Tensile Strength of Rock Materials. International Journal of Rock Mechanics and Mining Sciences & Geomechanics. Abst, (1978). , 15, 99-103.

[5] Thomée DIBWolski DGK, Plinninger RJ. The Modified Tension Test (MTT)-Evaluation and Testing Experiences with a New and Simple Direct Tension Test. Eurock (2004). Oct 23.

[6] Schmitt, D, & Zoback, M. Infiltration effects in the tensile rupture of thin walled cylinders of glass and granite: Implications for the hydraulic fracturing breakdown equation. International Journal of Rock Mechanics and Mining Sciences & Geomechanics Abstracts. Elsevier; (1993). , 30(3), 289-303.

[7] Selvadurai APSJenner L. Radial Flow Permeability Testing of an Argillaceous Limestone. Ground Water. (2013). , 51, 100-07.

[8] Fairhurst, C, & Hudson, J. A. Draft ISRM suggested method for the complete stress-strain curve for intact rock in uniaxial compression. Int J Rock Mech Min. Elsevier; (1999). , 36(3), 279-89.

[9] Zoback, M, Rummel, F, Jung, R, & Raleigh, C. Laboratory hydraulic fracturing experiments in intact and pre-fractured rock. International Journal of Rock Mechanics and Mining Sciences & Geomechanics Abstracts. Elsevier; (1977). , 14(2), 49-58.

[10] Plinninger, R. J, Wolski, K, Spaun, G, Thomée, B, & Schikora, K. Experimental and model studies on the Modified Tension Test (MTT)-a new and simple testing method for direct tension tests. GeoTechnical Measurements and Modelling-Karlsruhe. (2003). , 2003, 361-6.

Optimizing Stimulation of Fractured Reservoirs

Investigation of Hydraulic and Natural Fracture Interaction: Numerical Modeling or Artificial Intelligence?

Reza Keshavarzi and Reza Jahanbakhshi

Additional information is available at the end of the chapter

Abstract

Hydraulic fracturing of naturally fractured reservoirs is a critical issue for petroleum industry, as fractures can have complex growth patterns when propagating in systems of natural
fractures. Hydraulic and natural fracture interaction may lead to significant diversion of hydraulic fracture paths due to intersection with natural fractures which causes difficulties in
proppant transport and eventually job failure. In this study, a comparison has been made
between numerical modeling and artificial intelligence to investigate hydraulic and natural
frcature interaction. First of all an eXtended Finite Element Method (XFEM) model has been
developed to account for hydraulic fracture propagation and interaction with natural fracture in naturally fractured reservoirs including fractures intersection criteria into the model.
It is assumed that fractures are propagating in an elastic medium under plane strain and
quasi-static conditions. Comparison of the numerical and experimental studies results has
shown good agreement. Secondly, a feed-forward with back-propagation artificial neural
network approach has been developed to predict hydraulic fracture path (crossing/turning
into natural fracture) due to interaction with natural fracture based on experimental studies.
Effective parameters in hydraulic and natural fracture interaction such as in situ horizontal
differential stress, angle of approach, interfacial coefficient of friction, young's modulus of
the rock and flow rate of fracturing fluid are the inputs and hydraulic fracturing path(cross-
ing/turning into natural fracture) is the output of the developed artificial neural network.
The results have shown high potentiality of the developed artificial neural network approach to predict hydraulic fracturing path due to interaction with natural fracture. Finally,
both of the approaches have been examined by a set of experimental study data and the results have been compared. It is clearly observed that both of them yield promising results

while numerical modeling yields more detailed results which can be used for further investigations but it is computationally more expensive and time-consuming than artificial neural network approach. On the other hand, since artificial neural network approach is mainly data-driven if just the input data is available (even while fracturing) the hydraulic fracture path (crossing/turning into natural fracture) can be predicted real-time and at the same time that fracturing is happening.

1. Introduction

Hydraulic fracture growth through naturally fractured reservoirs presents theoretical, design, and application challenges since hydraulic and natural fracture interaction can significantly affect hydraulic fracturing propagation. Although hydraulic fracturing has been used for decades for the stimulation of oil and gas reservoirs, a thorough understanding of the interaction between induced hydraulic fractures and natural fractures is still lacking. This is a key challenge especially in unconventional reservoirs, because without natural fractures, it is not possible to recover hydrocarbons from these reservoirs. Meanwhile, natural fracture systems are important and should be considered for optimal stimulation. For naturally fractured formations under reservoir conditions, natural fractures are narrow apertures which are around 10^{-5} to 10^{-3} m wide and have high length/width ratios (>1000:1) [1].Typically natural fractures are partially or completely sealed but this does not mean that they can be ignored while designing well completion processes since they act as planes of weakness reactivated during hydraulic fracturing treatments that improves the efficiency of stimulation [2]. The problem of hydraulic and natural fracture interaction has been widely investigated both experimentally [3, 4, 5, 6, 7, 8] and numerically [9, 10, 11, 12, 13, 14, 15, 16, 17, 18]. Many field experiments also demonstrated that a propagating hydraulic fracture encountering natural fractures may lead to arrest of fracture propagation, fluid flow into natural fracture, creation of multiple fractures and fracture offsets [19, 20, 21, 22] which will result in a reduced fracture width. This reduction in hydraulic fracture width may cause proppant bridging and consequent premature blocking of proppant transport (so-called screenout) [23, 24] and finally treatment failure. Although various authors have provided fracture interaction criteria [4, 5, 25] determining the induced fracture growth path due to interaction with pre-existing fracture and getting a viewpoint about variable or variables which have a decisive impact on hydraulic fracturing propagation in naturally fractured reservoirs is still unclear and highly controversial. However, experimental studies have suggested that horizontal differential stress, angle of approach and treatment pressure are the parameters affecting hydraulic and natural fracture interaction [4, 5, 6] but a comprehensive analysis of how different parameters influence the fracture behavior has not been fully investigated to date. In this way, in order to assess the outcome of hydraulic fracture stimulation in naturally fractured reservoirs the following questions should be answered:

What is the direction of hydraulic fracture propagation?

How will the propagating hydraulic fracture interact with the natural fracture?

Will the advancing hydraulic fracture cross the natural fracture or will it turn into it?

For the purpose of this study, a 2D eXtended finite element method (XFEM) has been compared with a feed-forward with back-propagation artificial neural network approach to account for hydraulic and natural fracture interaction.

2. Interaction between hydraulic and natural fractures

The interaction between pre-existing natural fractures and the advancing hydraulic fracture is a key issue leading to complex fracture patterns. Large populations of natural fractures are sealed by precipitated cements (Figure 1) which are weakly bonded with mineralization that even if there is no porosity in the sealed fractures, they may still serve as weak paths for the growing hydraulic fractures [2].

Figure 1. A weakly bonded fracture cement in a shale sample [26].

In this way, experimental studies [4, 5, 6] suggested several possibilities that may occur during hydraulic and natural fractures interaction. Blanton [4] conducted some experiments on naturally fractured Devonian shale as well as blocks of hydrostone in which the angle of approach and horizontal differential stress were varied to analyze hydraulic and natural fracture interaction in various angles of approach and horizontal differential stresses. He concluded that any change in angle of approach and horizontal differential stress can affect hydraulic fracture propagation behavior when it encounters a natural fracture which will be referred to as opening, arresting and crossing. Warpinski and Teufel [5] investigated the effect of geologic discontinuities on hydraulic fracture propagation by conducting mineback experiments and laboratory studies on Coconino sandstone having pre-existing joints. They observed three modes of induced fracture propagation which were crossing, arrest by opening the joint and arrest by shear slippage of the joint with no dilation and fluid flow along the joint. In 2008 [6] some laboratory experiments were performed to investigate the interaction between hydraulic and natural fractures. They also observed three types of interactions between hydraulic and pre-existing fractures which were the same as Warpinski and Teufel's observations. The above referenced experimental studies have investigated the initial interaction between the induced fracture and the natural fracture, however, in reality may be the hydraulic fracture is arrested by natural fracture temporarily but with continued pumping of the fluid, the hydraulic fracture may cross (Figure 2) or turn into the natural fracture (Figure 3).

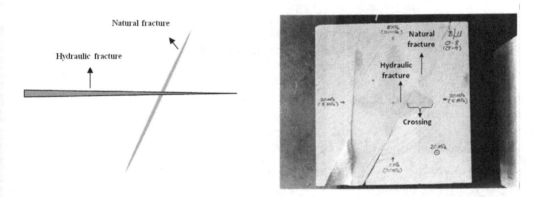

Figure 2. Propagating hydraulic fracture crosses the natural fracture and keep moving without any significant change in its path: left image is a schematic view of crossing and right image is the result of experimental study [4].

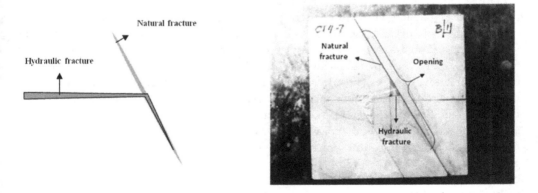

Figure 3. Hydraulic fracture turns into the natural fracture and propagates along it: left image is a schematic view and right image is the result of experimental study [4].

Alternatively, in some cases the hydraulic fracture may get arrested if the natural fracture is long enough and favorably oriented to accept and divert the fluid.

3. Numerical modeling: Extended Finite Element Method (XFEM)

For fracture propagation through numerical modeling an energy based criterion has been considered which is energy release rate, G. The energy release rate, G, is related to the stress intensity factors through Eq. 1 [27]:

$$G = \frac{1}{E'}(K_I^2 + K_{II}^2) \tag{1}$$

where $E' = E$ for plane stress (E is Young's modulus) and $E' = E/(1 - v^2)$ for plane strain (where v is the Poisson's ratio). Energy release rate has been calculated by the J integral using the domain integral approach [28] whereas J integral is equivalent to the definition of the fracture energy release rate, G, for linear elastic medium. If the G is greater than a critical value, G_c, the fracture will propagate critically. The direction of hydraulic fracture propagation will be calculated by Eq. 2 [29]:

$$\theta_c = 2\tan^{-1} \frac{1}{4} \left(\frac{K_I}{K_{II}} \pm \sqrt{\left(\frac{K_I}{K_{II}} \right)^2 + 8} \right) \tag{2}$$

During hydraulic and natural fracture interaction at the intersection point the hydraulic fracture has more than one path to follow which are crossing and turning into natural fracture. The most likely path is the one that has the maximum G. So, at the intersection point energy release rate is calculated for both crossing (G_{cross}) and turning into natural fracture (G_{turn}), and if (G_{turn}/G_{cross})>1 hydraulic fracture turns into natural fracture while if (G_{turn}/G_{cross})< 1 crossing takes place and hydraulic fracture crosses the natural fracture. To examine the proposed mechanism, eXtended Finite Element method (XFEM) was applied which was first introduced by Belytschko and Black [30] in order to avoid explicit modeling of discrete cracks by enhancing the basic finite element solution. In comparison to the classical finite element method, XFEM provides significant benefits in the numerical modeling of fracture propagation and it overcomes the difficulties of the conventional finite element method for fracture analysis, such as restriction in remeshing after fracture growth and being able to consider arbitrary varying geometry of fractures [12]. XFEM enhances the basic finite element solution through the use of enrichment functions which are the Heaviside function for elements that are completely cut by the crack and Westergaard-type asymptotic functions for elements containing crack-tips [27]. The displacement field for a point "x" inside the domain can be approximated based on the XFEM formulation as below [31]:

$$u^h(x) = \sum_{I \in N_{nu}} N_I(x) \left(u_I + \underbrace{H(x)a_I}_{I \in N_{ua}} + \underbrace{\sum_{\alpha-1}^{4} F_\alpha(x)b_I^\alpha}_{I \in N_{ub}} \right) \qquad (3)$$

Where N_I is the finite element shape function, u_I is the nodal displacement vector associated with the continuous part of the finite element solution, $H(x)$is the Heaviside enrichment function where it takes the value +1 above the crack and –1 below the crack, a_I is the nodal enriched degree of freedom vector associated with the Heaviside (discontinuous) function, $F_\alpha(x)$ is the near-tip enrichment function, b_I^α is the nodal enriched degree of freedom vector associated with the asymptotic crack-tip function, N_u is the set of all nodes in the domain, N_α is the subset of nodes enriched with the Heaviside function and N_b is the subset of nodes enriched with the near tip functions. At the intersection point, instead of Heaviside enrichment function, Junction function will be applied as shown in Figure 4 [32]. By all means, XFEM is well-suited for modeling hydraulic fracture propagation and diversion in the presence of natural fracture.

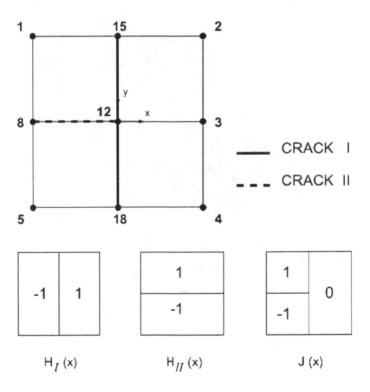

Figure 4. Definition of Junction function at the intersection point [32].

4. Artificial intelligence: Artificial Neural Network (ANN)

Artificial Neural Network (ANN) is considered as a different paradigm for computing and is being successfully applied across an extraordinary range of problem domains, in all areas of engineering. ANN is a non-linear mapping structure based on the function of the human brain that can solve complicated problems related to non-linear relations in various applications which makes it superior to conventional regression techniques [33, 34]. ANNs are capable of distinguishing complex patterns quickly with high accuracy without any assumptions about the nature and distribution of the data and they are not biased in their analysis. The most important aspect of ANNs is their capacity to realize the patterns in obscure and unknown data that are not perceptible to standard statistical methods. Statistical methods use ordinary models that need to add some terms to become flexible enough to satisfy experimental data, but ANNs are self-adaptable. The structure of the neural network is defined by the intercon-nection architecture between the neurons which are grouped into layers. A typical ANN mainly consists of an input layer, an output layer, and hidden layer(s) (Figure 5). As shown in Figure 5, each neuron of a layer is connected to each neuron of the next layer. Signals are passed between neurons over the connecting links. Each connecting link has an associated weight which multiplies by the related input.

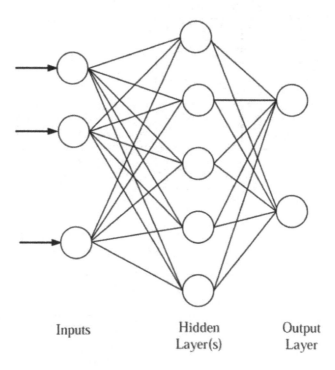

Inputs   Hidden Layer(s)   Output Layer

Figure 5. Schematic structure of an ANN.

Also, to diversify the various processing elements, a bias is added to the sum of weighted inputs called net input shown in Eq. (4)

$$n = \left(w_{S,1}P_1 + w_{S,2}P_2 + w_{S,3}P_3 + \dots + w_{S,R}P_R \right) + b_S \tag{4}$$

Where n is the net input, w is the weight, p is the input, b is the bias, S is the number of neurons in the current layer and R is the number of neurons in the previous layer.

Each neuron applies an activation function to its input to determine its output signal [35]. Neurons may use any differentiable activation function to generate their output based on problem requirement. The most useful activation functions are as follows:

$$a = purelin\left(n\right) \ = \ n \tag{5}$$

$$a \ = \ tansig\left(n\right) = \ \left(2 / \left(1 + exp\left(-2n\right)\right)\right) \ -1 \tag{6}$$

$$a = radbas(n) = \exp(-n^2) \tag{7}$$

where a is the neuron layer output. Purelin is a linear activation function (Figure 6A) defined in Eq. (5). Tansig is hyperbolic tangent sigmoid activation function (Figure 6B) mathematically shown in Eq. (6). Radbas is Gaussian activation function (Figure 6C) shown in Eq. (7). In Figure. 7, a one layer network with R inputs and S neurons is shown [36]. The optimum number of hidden layers and the number of neurons in these layers are determined by trial and error during the training/learning process. The hidden layers in the network are used to develop the relationship between the variables. In general, multilayer networks are more powerful than single-layer networks [37].

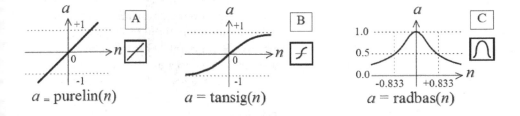

Figure 6. Common Activation functions.

Input   Layer of Neurons

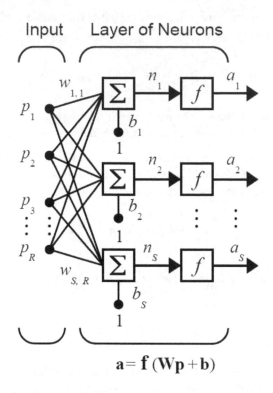

$$a = f(Wp + b)$$

Figure 7. A one layer network architecture with "R" inputs and "S" neurons.

4.1. Feed-forward network with back-propagation

The feed-forward network with back-propagation (FFBP) is one of the most eminent and widespread ANNs in engineering applications [38]. In addition, it is easy to implement and solves many types of problems correctly [39]. Usually, FFBP uses tansig and purelin as activation functions in the hidden and output layers, respectively and the net input is calculated the same as Eq. 4. FFBP operates in two steps. First, the phase in which the input information at the input nodes is propagated forward to compute the output information signal at the output layer. In other words, in this step the input data are presented to the input layer and the activation functions process the information through the layers until the network's response is generated at the output layer. Second, the phase in which adjustments to the connection strengths are made based on the differences between the computed and observed information signals at the output. In this step, the network's response is compared to the desired output and if it does not agree, an error is generated. The error signals are then transmitted back from the output layer to each node in the hidden layer(s) [40]. Then, based on the error signals received, connection weights between layer neurons and biases are updated. In this way, the network learns to reproduce outputs by learning patterns contained within the data. One iteration of this algorithm can be written as Eq. 8:

$$X_{k+1} = X_k - \alpha_k g_k \tag{8}$$

where X_k is a vector of current weights and biases, g_k is the current gradient, and α_k is the learning rate. Once the network is trained, it can then make predictions from a new set of inputs that was not used to train the network.

4.2. ANN performance criteria

There are several quantitative measures to assess ANN performance that the most usual one in a binary classification test is accuracy [41] (Fawcett 2006). To understand the meaning of accuracy, some definitions like true positive, false positive, true negative and false negative should be explained. Imagine a scenario where the occurrence of an event is considered. The test outcome can be positive (occurrence of the event) or negative (the event doesn't occur). According to this scenario:

- True Positive (TP): The event occurs and it is correctly diagnosed as it occurs;

- False Positive (FP): The event doesn't occur but it is incorrectly diagnosed as it occurs;

- True Negative (TN): The event doesn't occur and it is correctly diagnosed as it doesn't occur;

- False Negative (FN): The event occurs but it is incorrectly diagnosed as it doesn't occur.

According to above definitions:

$$\text{Accuracy} = \left(TP + TN\right) / \left(TP + TN + FP + FN\right) \tag{9}$$

In general, the accuracy of a system is a degree of closeness of the measured values to the actual (true) values [42].

5. Results and discussions

Physically, modeling hydraulic fracturing is a complicated phenomenon due to the heterogeneity of the earth structure, in-situ stresses, rock behavior and the physical complexities of the problem, hence if natural fractures are added up to the problem it gets much more complex in both field operation and numerical aspects. For simplicity, it is assumed that rock is a homogeneous isotropic material and the fractures are propagating in an elastic medium under plane strain and quasi-static conditions driven by a constant and uniform net pressure throughout the hydraulic fracture system. Fracturing fluid pressure is included in the model by putting force tractions on the necessary degrees of freedom along the fracture. A schematic illustration for the problem has been presented in Figure 8 which shows that hydraulic fracture propagates toward the natural fracture and intersects with it at a specific angle of approach, θ, and in-situ horizontal differential stress, $(\sigma_1 - \sigma_3)$.

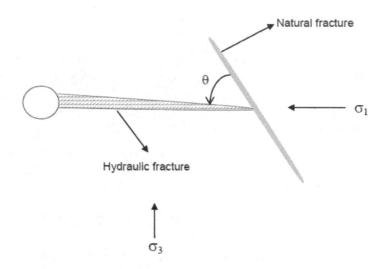

Figure 8. Schematic of hydraulic fracture intersecting pre-existing natural fracture [24].

So, a 2D XFEM code has been developed to model hydraulic fracture propagation in naturally fractured reservoirs and interaction with natural fractures. For this purpose, firstly Warpinski and Teufel's [5] experiments have been modeled to see how much the results of the developed XFEM model for hydraulic and natural fracture interaction, are compatible with them. Table 1, presents the results of XFEM code which can be compared with Warpinski and Teufel's [5] experiments. As shown in Table 1, the results of XFEM code indicate that at high to medium angles of approach, crossing and turning into natural fracture both are observed depending on the differential stress while at low angles of approach with low to high differential stress, the predominant case during hydraulic and natural fracture interaction is hydraulic fracture diversion along natural fracture which are in good agreement with Warpinski and Teufel's [5] experiments.

Angle of approach ($\theta°$)	Max. horizontal stress (psi)	min. horizontal stress (psi)	Horizontal differential stress (psi)	Experimental results [5]	G_{turn}/G_{cross}	XFEM results
30	1000	500	500	Turn into	3.46	Turn into
30	1500	500	1000	Turn into	2.05	Turn into
30	2000	500	1500	Turn into	1.29	Turn into
60	1000	500	500	Turn into	1.948	Turn into
60	1500	500	1000	Turn into	1.201	Turn into
60	2000	500	1500	Crossing	0.785	Crossing
90	1000	500	500	Turn into	1.013	Turn into
90	1500	500	1000	Crossing	0.833	Crossing
90	2000	500	1500	Crossing	0.598	Crossing

Table 1. Comparison of XFEM code results with Warpinski and Teufel's [5] experiments

Meanwhile debonding of natural fracture prior to hydraulic and natural fracture intersection could also be modeled which is a complicated and very interesting phenomena that has been rarely investigated. Figure 9, presents pre-existing fracture debonding before intersection with hydraulic fracture at approaching angles of 30°, 60°, 90° in Warpinski and Teufel's [5] experiments. As it is clearly observed in stress maps in Figure 9, a tensile stress is exerted ahead of hydraulic fracture tip for all of the approaching angles which makes the natural fracture debonded. In addition, the length and the position of the debonded zone vary depending on natural fracture orientation and horizontal differential stress.

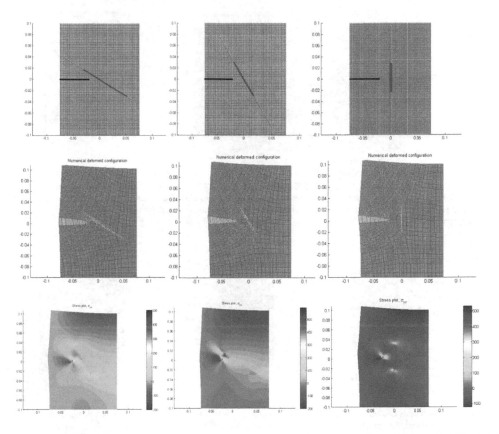

Figure 9. Natural fracture debonding before intersecting with hydraulic fracture at 30° (horizontal differential stress=1500 psi), 60° (horizontal differential stress=1000 psi), 90° (horizontal differential stress=1500 psi): the upper images show the coordinates of hydraulic and natural fracture relative to each other where the debonded zones are highlighted in red, the middle images the are the numerical deformed configurations (magnified by 3) and the images below them are the stress maps (σ_{xx}) (magnified by 3).

Figure 10, shows the debonded zone at the intersecting point of hydraulic and natural fracture and Figure 11 presents the result of hydraulic and natural fracture interaction for approaching angles of 30°, 60°, 90°.

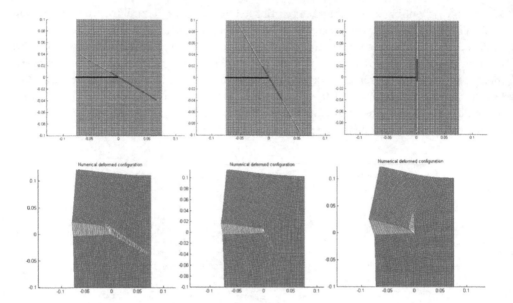

Figure 10. Debonded zones (highlighted in red) of natural fracture at the intersecting point with hydraulic fracture at 30° (horizontal differential stress=1500 psi), 60° (horizontal differential stress=1000 psi), 90° (horizontal differential stress=1500 psi): the upper images show the coordinates of hydraulic and natural fracture relative to each other where the debonded zones are highlighted in red and the images below them are the numerical deformed configurations (magnified by 3).

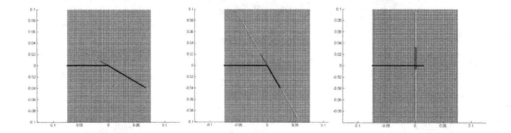

Figure 11. The results of hydraulic and natural fracture interaction after intersection: the left image is a natural fracture with the orientation of 30° (horizontal differential stress=1500 psi), the middle image is a natural fracture with the orientation of 60° (horizontal differential stress=1000 psi) and the right image shows a natural fracture with the orientation of 90° (horizontal differential stress=1500 psi).

In the second step, a FFBP neural network has been applied for predicting growing hydraulic fracturing path due to interaction with natural fracture in such a way that horizontal differential stress, angle of approach, interfacial coefficient of friction, young's modulus of the rock and flow rate of fracturing fluid are the inputs and hydraulic fracturing path (crossing or turning into natural fracture) is the output whereas tansig is an activation function. The data set used in this model consists of around 100 data based on experimental studies [4, 5, 6]. Table 2, represents the range of the parameters used in the developed ANN.

Parameter	Min	Max
Horizontal differential stress (psi)	290	2175
Angle of approach (deg)	30	90
Interfacial coefficient of friction	0.38	1.21
Young's modulus of the rock (psi)	$1.218*10^6$	$1.45*10^6$
Flow rate of fracturing fluid (m³/s)	$4.2*10^{-9}$	$8.2*10^{-7}$

Table 2. Range of the parameters used in the FFBP model.

The data in the database were randomly divided into two subsets. The training subset used 70% and the remaining 30%, was used for the testing subset. For standardizing the range of the input data and improves the training process the data used in network development were pre-processed by normalizing. Normalizing the data enhances the fairness of training by preventing an input with large values from swamping out another input that is equally important but with smaller values [43]. The optimal number of the neurons of a single hidden layer network for the developed ANN using trial and error method based on accuracy is 19 which is shown in Table 3. The developed FFBP neural network represented a high accuracy of 96.66% which was so promising. Also, according to the dataset, around 30 data were assigned for testing subset. Figure 12, show the results of the developed FFBP neural network predictions with actual measurements for testing subset.

Number of hidden neurons	Accuracy (%)
10	83.33
11	90
12	86.67
13	80
14	90
15	83.33
16	83.33
17	90
18	93.33
19	96.66
20	90

Table 3. Developed FFBP model designing with different neurons in the hidden layer.

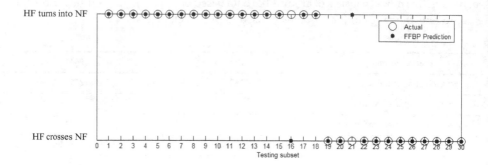

Figure 12. Comparison between actual case and FFBP prediction for testing subset.

As shown in Figure 12, FFBP predictions are in prominent agreement with actual measurements which shows the high efficiency of the developed FFBP neural network approach for predicting hydraulic fracturing path due to interaction with natural fracture based on horizontal differential stress, angle of approach, interfacial coefficient of friction, young's modulus of the rock and flow rate of fracturing fluid. Finally, both XFEM and ANN approaches have been examined by a set of experimental study data [4, 6] and the results have been compared. The results of a comparison are presented in Table 4. As shown in Table 4 both of the proposed approaches yield quite promising results and in just one case ANN approach result doesn't agree with the actual case.

Horizontal Differential Stress (psi)	Angle of approach (deg)	Coefficient of friction	E*10⁶ (psi)	Flow rate of fracturing fluid (m³/s)	Actual Case	XFEM Result	FFBP Prediction
290	60	0.6	1.45	8.2 e-7	T*	T	T
1885	45	0.6	1.45	8.2 e-7	T	T	T
1340	30	0.68	1.3	1 e-7	T	T	T
410	90	0.68	1.3	1 e-7	T	T	T
725	60	0.38	1.218	4 e-9	T	T	T
1015	30	0.38	1.218	4 e-9	T	T	T
913.5	60	0.38	1.218	4 e-9	T	T	C*
2175	60	0.6	1.45	8.2 e-7	C	C	C
1522.5	90	0.89	1.218	4 e-9	C	C	C
1595	90	0.89	1.218	4 e-9	C	C	C

* T= Turning into natural fracture

* C= Crossing natural fracture

Table 4. Comparison between actual case, XFEM results and FFBP prediction

6. Conclusions

Two new numerical modeling and artificial intelligence methodologies were introduced and compared to account for hydraulic and natural fracture interaction. First a new approach has been proposed through XFEM model and an energy criterion has been applied to predict hydraulic fracture path due to interaction with natural fracture. To validate and show the efficiency of the developed XFEM code, firstly the results obtained from XFEM model have been compared with experimental studies which shows good agreement. It's been concluded that natural fracture most probably will divert hydraulic fracture at low angles of approach while at high horizontal differential stress and angles of approach of 60 or greater, the hydraulic fracture crosses the natural fracture. Meanwhile, the growing hydraulic fracture exerts large tensile stress ahead of its tip which leads to debonding of sealed natural fracture before intersecting with hydraulic fracture that is a key point to demonstrate hydraulic and natural fracture behaviors before and after intersection. Then, a FFBP neural network was developed based on horizontal differential stress, angle of approach, interfacial coefficient of friction, young's modulus of the rock and flow rate of fracturing fluid and the ability and efficiency of the developed ANN approach to predict hydraulic fracturing path due to interaction with natural fracture was represented. The results indicate that the developed ANN is not only feasible but also yields quite accurate outcome. Finally, both of the approaches have been compared and both of them yield promising results. Numerical modeling yields more detailed results which can be used for further investigations and it can explain different observed behaviors of hydraulic fracturing in naturally fractured reservoirs as well as activation of natural fractures. Also, the potential conditions that may lead to hydraulic fracturing operation failure can be investigated through numerical modeling but it is computationally more expensive and time-consuming than artificial neural network approach. In another hand, since artificial neural network approach is mainly data-driven it can be of great use in real-time experimental studies and field hydraulic fracturing in naturally fractured reservoirs. So, as one may conclude easily, numerical modeling and artificial intelligence both have some positive and negative points; hence simultaneous use of these methods will lead to both technical and economical advantages in hydraulic fracturing operation especially in the presence of natural fractures.

Author details

Reza Keshavarzi and Reza Jahanbakhshi

Young Researchers and Elites Club, Science and Research Branch, Islamic Azad University, Tehran, Iran

References

[1] Liu, E. (2005). Effects of fracture aperture and roughness on hydraulic and mechanical properties of rocks: implication of seismic characterization of fractured reservoirs. Journal of Geophysics and Engineering, , 2, 38-47.

[2] Gale, J. F. W, Reed, R. M, & Holder, J. (2007). Natural fractures in the Barnett Shale and their importance for hydraulic fracture treatments, AAPG Bulletin, , 91, 603-622.

[3] Lamont, N, & Jessen, F. (1963). The Effects of Existing Fractures in Rocks on the Extension of Hydraulic Fractures. *JPT*, February., 203-209.

[4] Blanton, T. L. (1982). An Experimental Study of Interaction Between Hydraulically Induced and Pre-Existing Fractures. SPE 10847. Presented at SPE/DOE unconventional Gas Recovery Symposium, Pittsburg, Pennsylvania, May., 16-18.

[5] Warpinski, N. R, & Teufel, L. W. (1987). Influence of Geologic Discontinuities on Hydraulic Fracture Propagation. *JPT* February., 209-220.

[6] Zhou, J, Chen, M, Jin, Y, & Zhang, G. (2008). Analysis of fracture propagation behavior and fracture geometry using a tri-axial fracturing system in naturally fractured reservoirs. *International Journal of Rock Mechanics & Mining Sciences* (45) 1143-1152.

[7] Athavale, A. S, & Miskimins, J. L. (2008). Laboratory Hydraulic Fracturing Tests on Small Homogeneous and Laminated Blocks. 42nd US Rock Mechanics Symposium and 2nd U.S.-Canada Rock Mechanics Symposium, San Francisco, June July 2., 29.

[8] Zhou and Xue(2011). Experimental investigation of fracture interaction between natural fractures and hydraulic fracture in naturally fractured reservoirs. SPE EUROPEC/EAGE Annual Conference and Exhibition, Vienna, Austin, May , 23-26.

[9] Akulich, A. V, & Zvyagin, A. V. (2008). Interaction between Hydraulic and Natural Fractures. Fluid Dynamics. , 43, 428-435.

[10] Jeffrey, R. G, & Zhang, X. (2009). Hydraulic Fracture Offsetting in Naturally Fractured Reservoirs: Quantifying a Long-Recognized Process. SPE Hydraulic Fracturing Technology Conference, Woodlands, Texas, USA, January., 19-21.

[11] Rahman, M. M, Aghighi, A, & Rahman, S. S. (2009). Interaction between Induced Hydraulic Fracture and Pre-Existing Natural Fracture in a Poro-elastic Environment: Effect of Pore Pressure Change and the Orientation of Natural Fracture. SPE 122574. Presented at SPE Asia Pacific Oil and Gas Conference and Exhibition, Indonesia, August., 4-6.

[12] Dahi TaleghaniA. and Olson, J.E. (2009). Numerical Modeling of Multi-Stranded Hydraulic Fracture Propagation: Accounting for the Interaction between Induced and Natural Fractures. SPE 124884. Presented at SPE Annual Technical Conference and Exhibition, New Orleans, Louisiana, USA, October., 4-7.

[13] Chuprakov, D. A, Akulich, A. V, & Siebrits, E. (2010). Hydraulic Fracture Propagation in a Naturally Fractured Reservoir. SPE Oil and Gas India Conference, Mumbai, India, January., 20-22.

[14] Mclennan, J, Tran, D, Zhao, N, Thakur, S, Deo, M, Gil, I, & Damjanac, B. (2010). Modeling Fluid Invasion and Hydraulic Fracture Propagation in Naturally Fractured Rock: A Three-Dimensional Approach. SPE 127888, presented at International Symposium and Exhibition on Formation Damage Control held in Lafayette, Louisiana, USA, February., 10-12.

[15] Min, K. S, Zhang, Z, & Ghassemi, A. (2010). Numerical Analysis of Multiple Fracture Propagation in Heterogeneous Rock. 44th US Rock Mechanics Symposium and 5th U.S.-Canada Rock Mechanics Symposium, Salt Lake City, UT, June , 27-30.

[16] Keshavarzi, R, & Mohammadi, S. (2012). A New Approach for Numerical Modeling of Hydraulic Fracture Propagation in Naturally Fractured Reservoirs. SPE/EAGE European Unconventional Resources Conference and Exhibition Vienna, Austria, March., 20-22.

[17] Keshavarzi, R, Mohammadi, S, & Bayesteh, H. (2012). Hydraulic Fracture Propagation in Unconventional Reservoirs: The Role of Natural Fractures. 46th ARMA Symposium, Chicago, June., 24-27.

[18] Keshavarzi, R, & Jahanbakhshi, R. (2013). Real-Time Prediction of Complex Hydraulic Fracture Behaviour in Unconventional Naturally Fractured Reservoirs. SPE Middle East Unconventional Gas Conference and Exhibition, Muscat, Oman, January., 28-30.

[19] Stadulis, J. M. (1995). Development of a Completion Design to Control Screenouts Caused by Multiple Near-Wellbore Fractures. SPE 29549. Presented at Rocky Mountain Regional/Low Permeability Reservoirs Symposium and Exhibition, Denver, March., 19-22.

[20] Britt, L. K, & Hager, C. J. (1994). Hydraulic Fracturing in a Naturally Fractured Reservoir. SPE 28717. Presented at SPE International Petroleum Conference and Exhibition, Veracruz, Mexico, October., 10-13.

[21] Rodgerson, J. l. (2000). Impact of Natural Fractures in Hydraulic Fracturing of Tight Gas Sands. SPE 59540 Presented at SPE Permian Basin Oil and Gas Recovery Conference, Midland, Texas, March., 21-23.

[22] Jeffrey, R. G, Zhang, X, & Bunger, A. P. (2010). Hydraulic fracturing of naturally fractured reservoirs. Thirty-Fifth Workshop on Geothermal Reservoir Engineering Stanford University, Stanford, California, February , 1-3.

[23] Economides, M. J. (1995). *A practical companion to reservoir stimulation*. Elsevier Science Publishers. USA.

[24] Potluri, N, Zhu, D, & Hill, A. D. (2005). Effect of natural fractures on hydraulic fracture propagation. Presented at the SPE European formation damage Conference, Scheveningen, Netherlands, May., 25-27.

[25] Renshaw, C. E, & Pollard, D. D. (1995). An Experimentally Verified Criterion for Propagation across Unbonded Frictional Interfaces in Brittle, Linear Elastic Materials. *International Journal of Rock Mechanics Mining Science and Geomechanics*, (32) 237-249.

[26] Gale, J. F. W, & Laubach, S. (2010). Natural fracture study: Implications for development of effective drilling and completion technologies. The University of Texas at Austin, USA.

[27] Mohammadi, S. (2008). Extended finite element method for fracture analysis of structure. Blackwell Publishing, UK.

[28] Moran, B, & Shih, C. F. (1987). A general treatment of crack tip contour integrals. International Journal of Fracture, , 35, 295-310.

[29] Sukumar, N, & Prévost, J. H. (2003). Modeling quasi-static crack growth with the extended finite element method Part I: Computer implementation. International Journal of Solids and Structures, , 40, 7513-7537.

[30] Belytschko, T, & Black, T. (1999). Elastic crack growth in finite elements with minimal remeshing. International Journal of Fracture Mechanics, , 45, 601-620.

[31] Moës, N, Dolbow, J, & Belytschko, T. (1999). A finite element method for crack growth without remeshing. International Journal for Numerical Methods in Engineering , 46(1), 131-150.

[32] DauxCh., Moës, N., Dolbow, J., Sukumar, N. and Belytschko, T. (2000). International Journal for Numerical Methods in Engineering, , 48(12), 1741-1760.

[33] Sadiq, T, & Nashawi, I. S. (2000). Using Neural Networks for Prediction of Formation Fracture Gradient. SPE 65463. Presented at SPE/Petroleum Society of CIM International Conference on Horizontal Well Technology held in Calgary, Alberta, Canada, November., 6-8.

[34] Keshavarzi, R, Jahanbakhshi, R, & Rashidi, M. (2011). Predicting Formation Fracture Gradient in Oil and Gas Wells: A Neural Network Approach. 45th ARMA Symposium, San Francisco, June., 26-29.

[35] Mohaghegh, S. (2000). Virtual-Intelligence Applications in Petroleum Engineering, Parts Artificial Neural Networks. SPE 58046, *Distinguished Author Series*. doi:, 1.

[36] Demuth, H, & Beale, M. (1998). Neural *network toolbox for use with MATLAB*. User's Guide, Fifth Printing,Version 3. USA: Mathworks, Inc.

[37] Hagan, M. T, Demuth, H. B, & Beale, M. (1996). *Neural Network Design*. USA, Boston: PWS Publishing Company.

[38] Doraisamy, H, Ertekin, T, & Grader, A. S. (1998). Key Parameters Controlling the Performance of neuron- Simulation Applications in Field Development. SPE 51079. Presented at SPE eastern Regional Meeting, Pittsburgh, Pennsylvania, November., 9-11.

[39] Centilmen, A, Ertekin, T, & Grader, A. S. (1999). Applications of Neural-networks in Multi-well Field Development. SPE 56624. Presented at SPE Annual Technical Conference and Exhibition, Houston, Texas, October. doi:

[40] Ali, J. K. (1994). Neural Networks: A New Tool for the Petroleum Industry?.SPE 27561. Presented at European Petroleum Computer Conference held in Aberdeen, U.K., March., 15-17.

[41] Fawcett, T. (2006). *An introduction to ROC analysis.* Pattern Recognition Letters , 27, 861-874.

[42] Taylor, J. R. (1999). *An Introduction to Error Analysis: The Study of Uncertainties in Physical Measurements.* University Science Books. , 128-129.

[43] Al-fattah, S. M, & Startzman, R. A. (2001). Predicting Natural Gas Production Using Artificial Neural Network. SPE 68593. Presented at SPE Hydrocarbon Economics and Evaluation Symposium, Dallas, Texas, April., 2-3.

Chapter 11

Optimizing Hydraulic Fracturing Treatment Integrating Geomechanical Analysis and Reservoir Simulation for a Fractured Tight Gas Reservoir, Tarim Basin, China

Feng Gui, Khalil Rahman, Daniel Moos,
George Vassilellis, Chao Li, Qing Liu, Fuxiang Zhang,
Jianxin Peng, Xuefang Yuan and Guoqing Zou

Additional information is available at the end of the chapter

Abstract

A comprehensive geomechanical study was carried out to optimize stimulation for a fractured tight gas reservoir in the northwest Tarim Basin. Conventional gel fracturing and acidizing operations carried out in the field previously failed to yield the expected productivity. The objective of this study was to assess the effectiveness of slickwater or low-viscosity stimulation of natural fractures by shear slippage, creating a conductive, complex fracture network. This type of stimulation is proven to successfully exploit shale gas resources in many fields in the United States.

A field-scale geomechanical model was built using core, well log, drilling data and experiences characterizing the in-situ stress, pore pressure and rock mechanical properties in both overburden and reservoir sections. Borehole image data collected in three offset wells were used to characterize the in-situ natural fracture system in the reservoir. The pressure required to stimulate the natural fracture systems by shear slippage in the current stress field was predicted. The injection of low-viscosity slickwater was simulated and the resulting shape of the stimulated reservoir volume was predicted using a dual-porosity, dual-permeability finite-difference flow simulator with anisotropic, pressure-sensitive reservoir properties. A hydraulic fracturing design and evaluation simulator was used to model the geometry and conductivity of the principal hydraulic fracture filled with proppant. Fracture growth in the presence of the lithology-based stress contrast and rock properties was computed, taking into account leakage of the injected fluid into the stimulated reservoir volume

Optimizing Hydraulic Fracturing Treatment Integrating Geomechanical Analysis and Reservoir
Simulation for a Fractured Tight Gas Reservoir, Tarim Basin, China

179

predicted previously by reservoir simulation. It was found that four-stage fracturing was necessary to cover the entire reservoir thickness. Post-stimulation gas production was then predicted using the geometry and conductivity of the four propped fractures and the enhanced permeability in the simulated volume due to shear slippage of natural fractures, using a dual-porosity, dual-permeability reservoir simulator.

For the purpose of comparison, a conventional gel fracturing treatment was also designed for the same well. It was found that two-stage gel fracturing was sufficient to cover the whole reservoir thickness. The gas production profile including these two propped fractures was also estimated using the reservoir simulator.

The modeling comparison shows that the average gas flow rate after slickwater or low-viscosity treatment could be as much as three times greater than the rate after gel fracturing. It was therefore decided to conduct the slickwater treatment in the well. Due to some operational complexities, the full stage 1 slickwater treatment could not be executed in the bottom zone and treatments in the other three zones have not been completed. However, the post-treatment production test results are very promising. The lessons learned in the planning, design, execution and production stages are expected to be a valuable guide for future treatments in the same field and elsewhere.

1. Introduction

Following the success in exploiting shale gas resources by multi-stage hydraulic fracturing with slickwaters or low-viscosity fluid (i.e., linear gel) in horizontal wells in North America, there has been a lot of interest in applying this technique to other regions and other types of tight reservoirs. This is due in part to the fact that conventional gel fracturing treatments have been less successful in some naturally fractured reservoirs due to excessive unexpected fluid loss and proppant bridging in natural fractures, leading often to premature screen-outs. Additionally, the high-viscosity gel left inside the natural fractures causes the loss of virgin permeability of the reservoir in the case of inefficient gel breaking.

However, the challenge for doing this is that the physical mechanism responsible for this kind of stimulation is yet to be fully understood and a standard work flow for design and evaluation is yet to be developed. Furthermore, industry so far mainly relies on performance analogs to improve understanding of each shale play, and thus it usually takes years to advance up the learning curve for determining which factors best affect well production [1].

Currently, the general opinion on the mechanism leading to the success of waterfrac in shale gas reservoirs is that a complex fracture network is created by stimulation of pre-existing natural fractures. Although it is difficult to observe the processes acting during stimulation, microseismic imaging has enabled us to understand that both simple, planar fractures and complex fracture networks can be created in hydraulic fracture stimulations under different settings [2]. Fracture complexity is thought to be enhanced when pre-existing fractures are oriented at an angle to the maximum stress direction, or when both horizontal stresses and horizontal stress anisotropy are low, because these combinations of stress and natural fractures allow fractures in multiple orientations to be stimulated [3]. The result of stimulation therefore depends both on the geometry of the pre-existing fracture systems and on the in-situ stress state. It is now generally accepted that stimulation in shale gas reservoirs occurs through a combination of shear slip and opening of pre-existing (closed) fractures and the creation of new hydraulic (tensile) fractures [4-6]. In wells that are drilled along the minimum horizontal stress (S_{hmin}) direction, stimulation generally creates a primary radial hydraulic fracture that is perpendicular to S_{hmin}. Then, pressure changes caused by fluid diffusion into the surrounding rock and the modified near-fracture stress field induced by fracture opening cause shear slip on pre-existing natural fractures. If the horizontal stress difference is small enough, new hydraulic fractures perpendicular to the main fracture can open. Each slip or oblique opening event radiates seismic energy, which, if the event is large enough, can be detected using downhole or surface geophones.

Founded on the idea that productivity enhancement due to stimulation results not just from creation of new hydraulic fractures but also from the effect of the stimulation on pre-existing fractures (joints and small faults), a new workflow dubbed "shale engineering", was established by combining surface and downhole seismic, petro-physical, microseismic, stimulation, and production data [7, 8]. In this new workflow (Figure 1), the change in flow properties of natural fractures is predicted using a comprehensive geomechanical model based on the concept of critically stressed fractures [9-11]. Existing reservoir simulation tools can then be used to model the hysteresis of fracture flow properties that result from the microseismically detectable shear slip, which is critical to the permanent enhancement in flow properties and increased access to the reservoir that results from stimulation. The primary hydraulic fracture created and propped during the stimulation can be modeled using conventional commercial hydraulic fracture models by taking into account fluid leaked into natural fractures in the surrounding region. The propped conductivity is estimated using laboratory-based proppant conductivity data adjusted for the proppant concentration in the fracture. The propped main fracture model and the reservoir model with stimulated natural fracture properties can then be integrated into production simulators to predict production after the slickwater hydraulic fracturing treatment. When available, microseismic data can be used to help define the network of stimulated natural fractures that comprises the stimulated reservoir volume (SRV).

Although this new workflow was developed based on experiences in shale gas reservoirs, we believe it can also be applied to any unconventional reservoir requiring stimulation that has pre-existing natural fractures. Both Coal Bed Methane (CBM) and fractured tight gas reservoirs are examples of where this approach could be applied. In this paper, we will illustrate the workflow using the results of a study conducted in a fractured tight gas reservoir in the Kuqa Depression, Tarim Basin.

2. Project background

The project discussed in this paper was initiated to investigate various methods and practices to improve the economics of the field. Conventional gel fracturing had been tested in a few

Optimizing Hydraulic Fracturing Treatment Integrating Geomechanical Analysis and Reservoir
Simulation for a Fractured Tight Gas Reservoir, Tarim Basin, China

181

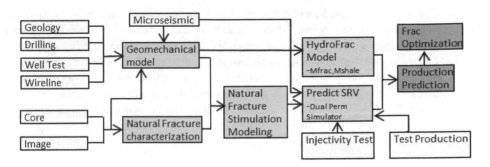

Figure 1. Workflow for predicting the complex fracture network developed by stimulating fractured reservoirs using low-viscosity fluid.

wells with disappointing results. One fault block (see Figure 2) was chosen as the target of a pilot study that included building a geomechanical model, optimizing hydraulic fracturing design, assessing the stability of faults near the target well, and (although it is not discussed here) analyzing wellbore stability for drilling horizontal wells. Three vertical wells were drilled. D2 and D3 are near the crest of the structure and D1 is ~ 2.5-3 km to the west. The main target is Cretaceous tight sandstone occurring at ~5300m to ~6000m depth. Reservoir rock is composed of fine sandstone and siltstones interlayered with thin shales. Average reservoir porosity is ~7% and average permeability is ~0.07 mD. The gross reservoir thickness is ~180-220m in this fault block. Wells D1 and D2 were completed by acidizing and gel fracturing; test production was ~15-27 ×10⁴m³/d. The objective of this project was to optimize hydraulic fracturing design for Well D3 based on the geomechanical analysis and investigate whether it is better to conduct slickwater treatment in the D3 well to stimulate and create a complex fracture network or utilize conventional two-wing gel fracturing.

Comprehensive datasets were available for all three wells including drilling experiences, wireline logs, image data, mini-fracs and well tests. Laboratory tests were also conducted on cores from well D2 to estimate the rock mechanical properties of reservoir rocks.

3. Geomechanical model

A geomechanical model includes a description of in-situ stresses and of rock mechanical and structural properties. The key components include three principal stresses (vertical stress (S_v), maximum horizontal stress (S_{Hmax}) and minimum horizontal stress (S_{hmin})), pore pressure (Pp) and rock mechanical properties, such as elastic properties, uniaxial compressive strength (UCS) and internal friction. The relative magnitude of the three principal stresses and the consequent orientation of the most likely slipping fault or fracture define the stress regime to be normal faulting ($S_v>S_{Hmax}>S_{hmin}$), strike-slip faulting ($S_{Hmax}>S_v>S_{hmin}$) or reverse faulting ($S_{Hmax}>S_{hmin}>S_v$). The horizontal stresses are highest relative to the vertical stress in a reverse faulting regime and lowest relative to the vertical stress in a normal faulting regime. Hydraulic fractures are vertical and propagate in the direction of the greatest horizontal stress in a strike-slip or normal faulting regime. In a reverse faulting stress regime in which S_v is the minimum stress, hydrofractures are horizontal. These different stress regimes also have consequences for the pressure that is required to open a network of orthogonal hydrofractures by stimulation. In places where the horizontal stresses are low and nearly equal, a relatively small excess pressure above the least stress may be required to open orthogonal fractures. Where the horizontal stress difference is larger, a larger excess pressure is required to open orthogonal fractures. Where the least stress is only slightly less than the vertical stress, weak horizontal bed boundaries and mechanical properties contrasts between layers may allow opening during stimulation of horizontal bedding ("T-fractures").

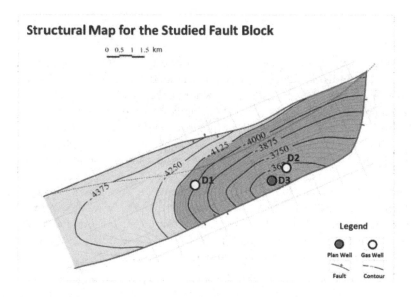

Figure 2. Structural map showing the offset well locations

Except for the magnitude of S_{Hmax}, other components of the geomechanical model can be determined using borehole data by reviewing a few representative wells in the field. Vertical stress is calculated by integrating formation density, which is obtained from wireline logs. The magnitude of S_v across this fault block is in a similar range. Pore pressure was constrained, mainly by referencing direct measurement data and drilling experiences. This is due to the complex tectonic history. Conventional under-compaction approaches for pore pressure estimation may not apply in the study area. Evidence for this is the over-compacted density profile. In addition, due to the complex lithology changes the log response with depth may reflect lithology changes rather than pressure variation. Well test data from D1 and D2 showed that the reservoir pressure is ~88-90 MPa, an equivalent pressure gradient of ~1.6-1.7 SG, which is abnormally over-pressured.

Rock mechanical laboratory tests were conducted on cores from the sandstone reservoirs and the interlayered shales in the D2 well, and the results were used to constrain a log-calibrated

Optimizing Hydraulic Fracturing Treatment Integrating Geomechanical Analysis and Reservoir
Simulation for a Fractured Tight Gas Reservoir, Tarim Basin, China

183

range of UCS and other rock mechanical parameters. Figure 3 shows the match between log-derived rock strength profiles and laboratory test results in D2. Dynamic Young's modulus was calculated from compressional and shear velocities and density and calibrated to static values using laboratory test results. The relationship between dynamic and static Poisson's Ratio was not obvious; the dynamic Poisson's Ratio computed from Vp/Vs matched reasonably well with the laboratory results, so it was used directly in the modeling. Young's Modulus-based empirical relationships were used to estimate the UCS for both sandstone and inter-layered shales.

Minimum horizontal stress (S_{hmin}) at depth can be directly estimated from extended leak-off tests (XLOT), leak-off tests (LOT) or mini-frac tests. No extended leak-off tests were conducted in the field. LOTs and leak-off points from two reliable LOTs were used to constrain the upper limit of S_{hmin} (~2.09 SG EMW at ~4000 m TVD). One mini-frac test was conducted in the sandstone reservoir in D2, with the interpreted fracture closure pressure (closest estimation to S_{hmin}) ~2.064 ppg EMW at ~5400 m TVD. Because LOTs are usually conducted in shaly formations while mini-frac tests are usually carried out in sandstone reservoirs, the LOTs and mini-frac tests are used to construct separate S_{hmin} profiles in shales and sandstones, respectively using the effective stress ratio method (S_{hmin}-Pp/S_v-Pp). The effective stress ratio from LOT is ~0.725 and from mini-frac test is ~0.48, which indicates there is a dramatic stress difference between sandstones and shales (stress contrast). The contrast between different lithology significantly influences hydraulic fracturing design. The relative lower stress in sandstones indicates that a hydraulic fracture should be easily created in the tight sandstone, however, the interlayered shales which have higher stress act as frac barriers and pinch points, thereby complicating fracture propagation and the final fracture geometry and conductivity.

The azimuth and magnitude of maximum horizontal stress (S_{Hmax}) can be constrained through the analysis of wellbore failures such as breakouts and tensile cracks observed on wellbore images or multi-arm caliper data. Wellbore failure analysis allows constraining of the orientation and magnitude of the S_{Hmax} because stress-induced wellbore failures occur due to the stress concentration acting around the wellbore once is drilled. The presence, orientation, and severity of failure are a function of the in-situ stress fields, wellbore orientation, wellbore and formation pressures and rock strength [12]. High-resolution electrical wireline image logs were available in all three study wells. Both breakouts and drilling-induced tensile fractures (DITFs) were observed in the reservoir sections in D2 and D3 wells. Only DITFs were observed in well D1, which could be due to the higher mud weights used during drilling and the poor quality of the image data in lower part of the reservoir.

Figure 4 shows examples of the breakouts seen in the D3 well. The example shows the typical appearance of breakouts observed on images. Here, the average apparent break-out width is ~30-40 degree. The breakouts mostly occur in shales and more breakouts are observed in the lower part of reservoir where the formations become more shaly. The orientation of breakouts is quite consistent with depth and across the block. However, small fluctuations of breakout orientation can be observed locally while intercepting small faults (an example can be seen in the right plot in Figure 4). This may indicate that some of these

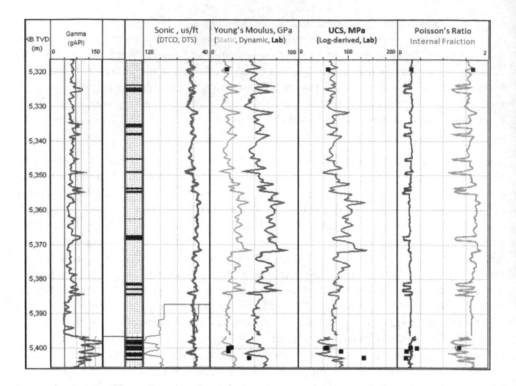

Figure 3. Comparison of laboratory (black squares) and log derived rock mechanical properties in D2 well

faults are close to or at the stage of being critically stressed. This has important implications for the stress state in the area and the likelihood of stimulating fractures by injection. Breakouts usually develop at the orientation of S_{hmin} and DITFs in the direction of S_{Hmax} in vertical and near-vertical wells. In the left plot of Figure 4, DITFs can also be observed in the same interval as the breakouts with an orientation that is ~90 degrees from the breakout directions, consistent with this expectation. DITFs are seen more often in sandstone than in shale. Based on wellbore breakouts and DITFs interpreted from the image data in D3, the azimuth of S_{Hmax} is inferred to be ~143° ±10 °. This is similar to the azimuth of S_{Hmax} inferred from wellbore failures observed in the other two wells. It is also consistent with the regional stress orientation from the World Stress Map [13].

The magnitude of S_{Hmax} is constrained by forward-modeling the stress conditions that are consistent with observations of wellbore failures observed on image logs, given the data on rock strength, pore pressure, minimum horizontal stress, vertical stress, and mud weight used to drill the well. Figure 5 is a crossplot of the magnitude of S_{hmin} and the magnitude of S_{Hmax}, which summarizes the results of S_{Hmax} modeling in D3. The magnitude of S_v (~2.49 SG) is indicated by the open circle. The modeling was conducted in both sandstone and shale. The rectangles in different colors are the possible S_{hmin} and S_{Hmax} ranges at every modeling depth. Modeling shows slightly different results for the S_{Hmax} and S_{hmin}

Optimizing Hydraulic Fracturing Treatment Integrating Geomechanical Analysis and Reservoir
Simulation for a Fractured Tight Gas Reservoir, Tarim Basin, China

185

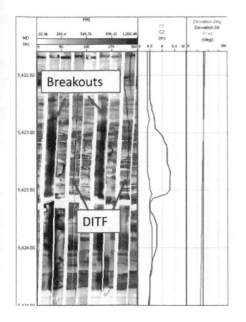
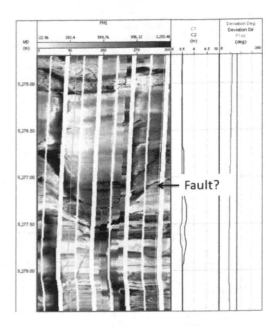

Figure 4. Drilling induced wellbore failures (breakouts & tensile fractures) observed on electrical image in D3 well.

magnitudes in the different lithologies. However, both results are consistent with the magnitudes of S_{hmin} inferred from LOTs and mini-fracs. Figure 5 shows that the magnitude of maximum horizontal stress is higher than the vertical stress in both cases, and higher in the shale than in the sand. Thus, the study area is in a strike-slip faulting stress regime ($S_{hmin} < S_v < S_{Hmax}$). The difference between the magnitudes of S_{Hmax} and S_{hmin} is ~0.8 SG in the reservoir section, suggesting high horizontal stress anisotropy. In such a condition, it is unlikely to open the natural fractures by tensile mode. However, the natural fractures might dilate in shear mode depending on their orientations and stress conditions. The final geomechanical model was verified by matching the predicted wellbore failure in these wells with that observed from image data and drilling experiences.

4. Natural fractures characterization and stimulation modeling

Natural fractures have been observed on cores and image logs in the study area. The fluid losses during drilling not only suggest the existence of natural fractures but also that some at least of these fractures are permeable in-situ. Based on the core photos shown in Figure 6, open high-angle tectonic fractures can be seen on cores from D2 and D3 wells near the crest of the structure. A fracture network consisting of a group of fractures with different orientations can be seen on the cores from the D1 well, and these fractures appear to have less apertures than high-angle fractures observed in D2 and D3.

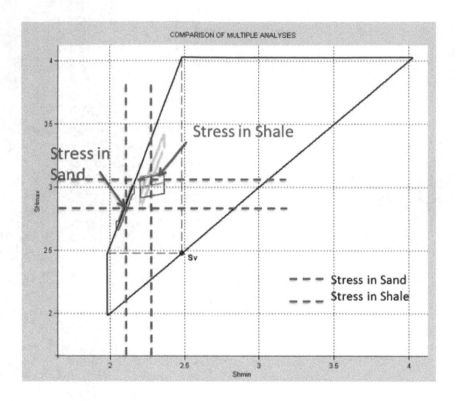

Figure 5. Stress modeling results in well D3. The range of horizontal stress magnitudes are consistent with the occurrence of wellbore failures (breakouts and DITFs) observed on wellbore images.

Natural fractures were interpreted and classified using high-resolution electrical images in all three wells. Based on the appearance on image data, the natural fractures are classified as below:

- Conductive: dark highly dipping planes on image logs

- Resistive: white dipping planes on image log

- Critically Stressed: related to local failure rotation

- Fault: features discontinued across the dipping planes

- Drilling Enhanced: discontinuous and fracture traces are 180 degrees apart and in the direction in tension

Optimizing Hydraulic Fracturing Treatment Integrating Geomechanical Analysis and Reservoir
Simulation for a Fractured Tight Gas Reservoir, Tarim Basin, China

187

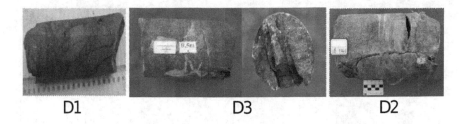

D1   D3   D2

Figure 6. Core photos showing the natural fractures observed in three offset wells

Figure 7 shows a few examples of natural fractures observed on the electrical images. The plot on the left shows some examples of high-angle and low-angle conductive fractures that appear to be continuous dark lines on the images. Flexible sinusoids can be fit to the fracture traces and fracture orientation can be determined. The plot on the right shows an example of drilling enhanced natural fractures for which the fracture trace is discontinuous. The fact that parts of these fractures can be detected on the electrical image is due to fluid penetration into the fracture at the orientation around where the rock at the borehole wall is in tension during drilling. The classification of the natural fractures indicates the relative strength of the fractures. For example, the resistive fractures are closed and mineralized. Active faults or critically stressed natural fractures might be open and conductive, even under the original conditions. During stimulation, these fractures are the most easily stimulated. However, it is important to note that the classification of natural fractures is purely based on their appearance on the electrical images, and cannot be used directly to quantify permeability or other flow properties.

Figure 8 shows the fractures orientations on a crossplot of the strike and dip angles of all fractures observed in the three wells. The natural fractures observed can be divided into three groups. The first group is low-angle fractures (dip<20 °), which could be related to beddings. The second group is the major fractures seen in this block that have intermediate dip angles (~25-55°) and strike at an azimuth of ~155°N. The third group consists of fractures with strikes of ~355°N and ~100°N and dip angles ~ 35°-65° and 25°-35°, respectively. Because of their wide range of orientations and cross-cutting relationship, these three groups of fractures could be stimulated to form a well-connected grid with a major fracture azimuth (~155°N) aligned with the direction of maximum horizontal stress (~143°N). This direction is nearly perpendicular to the faults, defining the shape of fault block (see Figure 2). Because the structural trends and the stresses are aligned, it enabled us to create a reservoir model with a grid that is consistent with both.

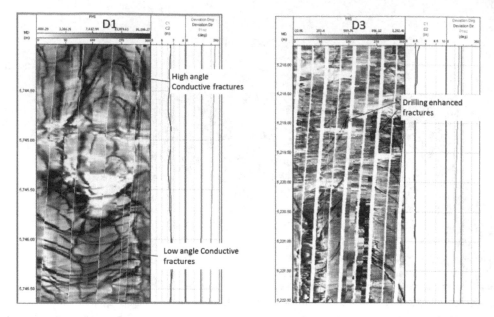

Figure 7. Examples of natural fractures observed on electrical image in D1 and D3 wells.

Effective stresses in the earth are always compressive, and natural processes tend to "heal" fractures through vein filling and other processes. Therefore, the intrinsic fracture aperture of most fractures is likely to be very small or even zero (cases where dissolution creates voids that prevent full closure are a notable exception). Thus, it is increasingly recognized that active processes are necessary to maintain fracture permeability. One such process is periodic slip along fractures that are critically stressed (i.e., those that are at or near the limiting ratio of shear to normal stress to slip). This process, and the influence of effective normal stresses on fracture aperture, can be modeled using a simple equation that describes the variation in aperture as a function of normal stress for a pure Mode I fracture. The same equation with different parameters can also be used to model the same fracture after slip has occurred [9-11].

$$a = \frac{A \bullet a_0}{(1 + 9\sigma'_n / B)} \tag{1}$$

Equation 1 is one example that describes aperture in terms of an initial aperture ($A \bullet a_0$) and an effective normal stress at which the aperture is only 10% as large (B). A and B both increase due to slip, resulting in a larger "unstressed" aperture and a stiffer fracture caused by "self-propping" due to generation during slip of a mismatch in the fracture faces and/or creation of minor amounts of rubble at the fracture face.

The contribution of fractures to the relative productivity of a well of any orientation can be computed by summing the contributions of all fractures, weighted by the product of their relative transmissivity (which is a function of aperture) and the likelihood of the well inter-

Optimizing Hydraulic Fracturing Treatment Integrating Geomechanical Analysis and Reservoir
Simulation for a Fractured Tight Gas Reservoir, Tarim Basin, China

189

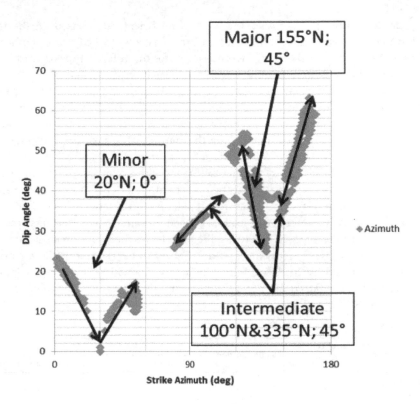

Figure 8. Cross-plot between the strike and dip angles of all the fractures observed in three offset wells.

secting the fracture (which is a function of the difference between the fracture and the well orientation). This relative productivity can be written as [10]

$$P_{well} = \Sigma_{fracs} \{ \max (|\hat{w} \bullet \hat{n}_i|, a) \times P_i \} \qquad (2)$$

where $\hat{w}$ and $\hat{n}_i$ are unit vectors along the axis of the well and normal to the ith fracture, a is a number representing the likelihood of a well intersecting a fracture if it lies in the plane of the fracture, and P_i is the relative permeability of the fracture.

The fractures interpreted from image data are only those that intersect the logged wells that are a function of their orientations, and there is no information about the fracture distribution between the wells. To ensure the most meaningful representation of the fractures in the reservoir, the fractures interpreted from all three wells were combined and the distribution was corrected to account for the likelihood of each fracture intersecting the well at the point where it was observed. This combined fracture data set was then used to model the productivities of wells in their natural condition and the change in productivity due to the shear-slip of natural fractures.

Figure 9 shows relative productivity for wells of all orientations based on the fractures observed in all three wells. Natural fractures are shown as poles to the fracture surfaces (black dots). Different apertures and strengths were assumed for the different types of fractures based on their classifications described above (Table 1). The plot on the left shows the relative productivity under pre-stimulation conditions, while the plot on the right shows the relative productivity calculated using equation 2 after the fractures were stimulated with a pressure 20 MPa above the original reservoir pressure. It can be seen that the maximum productivity increases by a factor of 5 if all fractures see the same 20-MPa pressure increase, which is obviously not the case during real stimulation. Superimposed on Figure 9 are the computed optimal orientations of wells based on the fracture and stress analysis (green circles). If none of the fractures is critically stressed, then the best orientation to drill a well is perpendicular to the largest population of natural fractures. If some fractures have enhanced permeability because they are critically stressed, the optimal orientation shifts in the direction of the greatest concentration of critically stressed fractures. Figure 9 shows there are some fractures already near or being critically stressed, even under ambient condition (left plot), and the maximum productivity is achieved by drilling highly deviated wells with ~20 °N hole azimuth. The optimum wellbore orientation after ~ 20-MPa stimulation is nearly horizontal and in the direction of ~228 °N.

Fracture classification	Fracture cohesion (MPa)	Sliding Friction	a_0	A		B (MPa)	
				Un-stimulated	Stimulated	Un-stimulated	Stimulated
Conductive	5	0.6	10	0.18	0.18	10	100
Resistive	5	0.6	10	0.1	0.18	1	100
Faults	0	0.6	30	0.18	0.18	100	100
Drilling enhanced	0	0.6	10	0.1	0.18	10	100
Critically Stressed	1	0.2	10	0.1	0.2	10	100

Table 1. Model parameters to calculate relative productivities for different types of natural fractures

Figure 10 shows the general effect of reservoir flow properties changes due to the natural fracture stimulation for studied fault block. Again, all the fractures interpreted from image logs in the three wells are used for modeling. Cross-plots between relative productivity (flow rate/pressure) vs. reservoir pressure are shown for three different cases: under original conditions, after 30-MPa and after 50-MPa stimulation. The blue curves show productivity changes during stimulation when the pressure is increasing, the green curves show the productivity changes during flowback and production. Modeling ends at ~20-MPa depletion. The relative productivity at ~20-MPa depletion increases five-fold after the 30-MPa stimulation (productivity increases from ~4 to ~20). There is no obvious improvement in the relative productivity of natural fractures for 50-MPa stimulation (bottom left) compared to 30-MPa stimulation. The bottom-right plot shows the number of stimulated natural fractures under different pressure conditions. It is clear that nearly all of the natural fractures are stimulated while the pressure increases to ~130 MPa (40-Ma stimulation), which explains why there is

Optimizing Hydraulic Fracturing Treatment Integrating Geomechanical Analysis and Reservoir
Simulation for a Fractured Tight Gas Reservoir, Tarim Basin, China

191

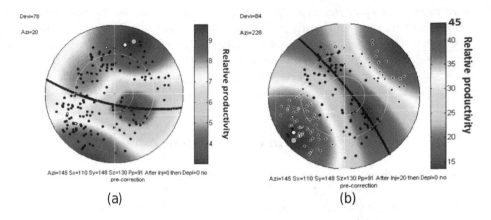

Figure 9. Relative well productivity for wells of all orientations based on the fractures observed in all three wells. (a) Ambient condition. (b) After 20-MPa stimulation. Natural fractures are shown as poles to the fracture surfaces (black dots). Green circles are computed optimal orientations of wells with highest productivity from natural fractures based on the fracture and stress analysis.

little improvement with further stimulation. It is important to note that this result does not take into account the possibility of injecting proppant to maintain the conductivity of fractures which open at pressures above 40 MPa.

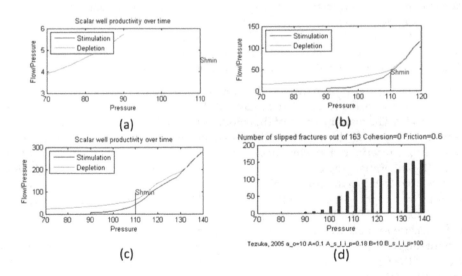

Figure 10. Reservoir flow properties changes with time due to natural fracture stimulation for studied fault block. The blue curves are showing the productivity changes when pressure increases during stimulation, the green curves are showing productivity changes during flowback and production. (a) no stimulation (b) 30-MPa stimulation (c) 50-MPa stimulation (d) number of stimulated (shear slip) natural fractures. Fracture properties: cohesion=0, sliding friction=0.6.

The above relative productivity modeling of natural fractures shows the conductivity of natural fractures increases significantly if the stimulation pressure is at or above the minimum horizontal stress. This is because many of the natural fractures are non-optimally oriented. Assuming a connected fracture network exists, the conductivity increase could be a factor of five for the stimulated fracture network while stimulation pressure is ~130MPa or higher (assuming the pressure reaches all fractures).

5. Predicting the shape of the stimulated reservoir volume

Fracture stimulation modeling showed that the shear slip of natural fractures could be effective in improving reservoir properties. Next, we need to reproduce the affected productive volume in the reservoir using the "shear stimulation" concept to enable more accurate production prediction. At the present no commercial simulator can fully model this process in 3D, although some research simulators have been developed. It was decided to use two different commercial models to simulate both fracture network stimulation created by low-viscosity frac fluid and the growth of the main hydraulic fracture. A commercial dual-porosity, dual-permeability simulator is used to simulate the flow property changes of natural fractures due to the shear slip. A commercial hydraulic fracturing design and evaluation simulator is used to model the geometry and conductivity of the principal hydraulic fracture filled with proppant. The modeling in two separate simulators is coupled by the fluid volume used for stimulation. The fluid volume leaked off in the shear-dilated natural fracture network was estimated in the dual-permeability, dual-porosity flow simulator. By adjusting the pressure-dependent leak-off coefficient, the fluid volume leaked off in the hydraulic fracturing simulator was matched with the fluid volume leaked into natural fractures networks estimated by the flow simulator. The prediction of the stimulated reservoir volume is discussed in the rest of this section and the hydraulic fracturing design will be discussed in next section.

To predict the extent and properties of the stimulated volume by a dual-permeability, dual-porosity simulator, a finely gridded model (Model A) was created based on the original reservoir model. The main function of this model is to simulate the change in flow properties in every single frac stage during and immediately after injection. The model is initialized with average known reservoir characteristics such as matrix porosity and permeability, fracture permeability and initial pressure, characterized from core and log analysis. Although different cases have been tested in the study, only one of the most realistic cases will be discussed here: the average matrix porosity used in the initial model is ~7.4%, matrix permeability is 0.07 mD in all directions, and the initial fracture permeability is ~ 0.2 mD. The initial fracture permeability is set close to the lower bound of fracture permeability based on core and log analysis. The orientations of the principal flow directions were chosen to correspond to the principal directions of the fracture sets and of bedding, which also approximately corresponded to the principal stress directions.

The relative magnitudes of the permeability enhancements in different directions were constrained by the geomechanical analysis. A set of permeability-pressure tables for different directions were then used to describe the hysteretic rock behavior that results from shear fracture activation. Although the fracture properties during stimulation can be estimated as described in the previous section, it is better to calibrate and constrain the permeability-pressure relationship based on real lab or in-situ tests, e.g., using a pre-stimulation injectivity test [4]. The injectivity test should ideally be conducted in the open hole using slow injection to evaluate the potential natural fractures being stimulated, as permeability changes could then be interpreted based on the flow-rate/pressure changes along with the reservoir pressure. Because the D3 well has already been cased it was impossible to conduct such a test in the field before the actual treatment is carried out. Consequently, it was decided to produce a permeability-pressure table based on experience from shale gas reservoirs. Based on this table, on fracture density in different directions and on the stress anisotropy, a composite transmissibility multiplier was produced for the prediction of properties and extent of the stimulated reservoir volume. Transmissibility multipliers were different for each of the I, J and K directions; those directions were aligned as discussed above with the primary structural fabric and stresses. The propagation of the pressure and fluid front in these directions can be controlled by modifying these multipliers.

Figure 11 shows diagrammatically the relationship between the permeability multiplier and the pore pressure (green curve). A slow increase in the permeability multiplier with increasing pressure occurs until fractures begin to slip. Above this pressure, the injectivity increases rapidly as an increased number of fractures are stimulated. During decreasing injection pressure in the injectivity test, the injectivity should decrease more slowly, retaining behind a permanent injectivity increase. The post-stimulation response can also be extrapolated to pressures below the original reservoir pressure. This makes it possible to predict the reservoir's response to depletion, which could lead to improved predictions of production decline. When the pressure during stimulation exceeds the minimum horizontal stress, extensional hydrofracs are created, and the permeability-pressure relationship does not follow the green line. Three different flow paths (A, B, C) were assumed for conditions with pressure above S_{hmin}, and the intermediate path, B was chosen to be used in the simulation.

The result of this modeling work is a 3D induced permeability map that describes the stimulated rock volume as discrete blocks, each with a unique permeability. The stimulated rock volume is therefore described not as a geometrical shape with identical flow properties throughout, but as a rock body with variable induced permeability, as shown in Figure 12.

6. Hydraulic fracturing design and reservoir simulation

As discussed earlier, a commercial simulator was used to model the hydraulic fracture created during the stimulation along with the stimulated natural fracture network using low-viscosity fluids. Stress profiles and other elastic rock properties estimated in the geomechanical analysis were used as input for the design. To achieve better proppant distribution, a low-viscosity linear gel was combined with slickwater in the treatment. The low-viscosity linear gel was optimized using different concentrations of ingredients for the high reservoir temperature (~126°C) using source water and local ingredients. Due to the high closure pressure and low viscosity of the fluid, high-strength small-mesh proppants were used in the design.

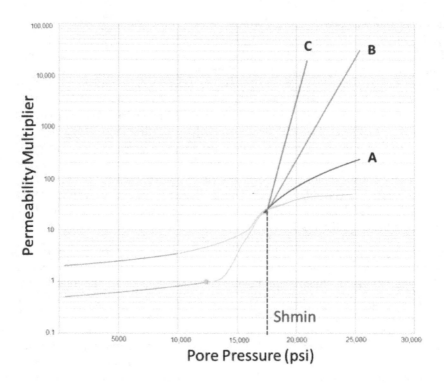

Figure 11. Relationship between the permeability multiplier and the pore pressure (green curve) for natural fractures used in the simulation. Three different flow paths (A, B, C) were assumed for conditions with pressure above Shmin.

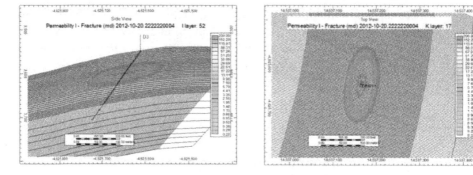

Figure 12. Side view (left) and top view (right) of the predicted 3D permeability map. The property shown in the plots is present fracture permeability.

Modeling showed that four stages would be required for slickwater/linear gel treatment to cover the 160 m thick reservoir due to the high leak off of low-viscosity fluids (Figure 13). A reasonable proppant distribution was achieved by using the low-viscosity linear gel.

Optimizing Hydraulic Fracturing Treatment Integrating Geomechanical Analysis and Reservoir
Simulation for a Fractured Tight Gas Reservoir, Tarim Basin, China

195

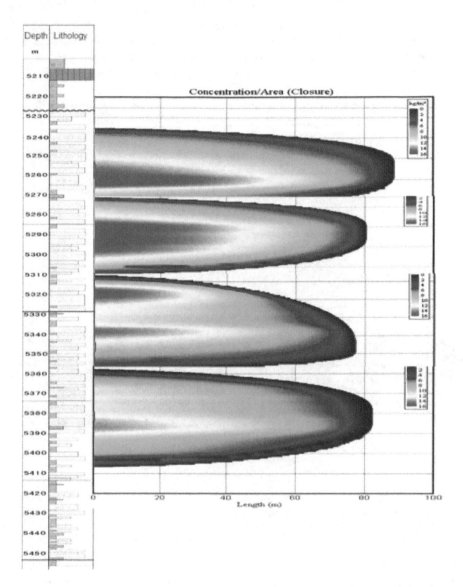

Figure 13. Fracture growth and proppant coverage (colour) for four stage of hydraulic fractures using slickwater/linear gel.

To predict the production after the stimulation, the propped hydraulic fractures were imported into the reservoir model with flow properties enhanced by stimulated natural fractures (Model A). Because the natural fracture distribution between wells is unknown, the same stimulated Model A was used for all four stages. The left plot of Figure 14 shows a side view of the reservoir model combining four Model A's with stimulated reservoir volumes and four propped hydraulic fractures, which was used for production prediction.

To compare the prediction result from slickwater/liner gel treatment with conventional gel fracturing, a conventional bi-wing hydraulic fracturing design using a high-viscosity gel was also developed. The gel fluid was optimized using different concentrations of ingredients for the high reservoir temperature (~126°C) using source water and local ingredients. The same type of proppant used for the slickwater/liner gel treatment was used for the design of gel treatment. The proppant concentrations and amounts will be certainly different in these two types of treatments. It was found that two stages were enough to cover the whole reservoir interval (Figure 15). These two designed hydraulic fractures were then imported into the original reservoir model (right plot in Figure 14) for production prediction and comparison of the production to that predicted after slickwater linear gel stimulation.

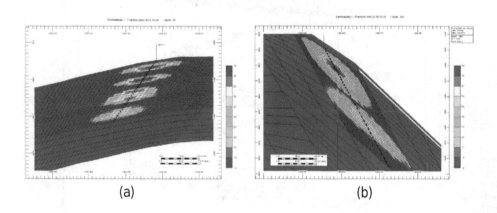

(a)   (b)

Figure 14. Side views of reservoir models showing fracture permeability used for production prediction. (a) Reservoir model combining four Model A's with stimulated reservoir volumes and four propped hydraulic fractures using slick-water and linear gel; (b) Original reservoir model and two propped hydraulic fractures using high-viscosity gel fluid.

Figure 16 shows the production prediction comparison from the two different hydraulic fracturing treatments. The red curve is the production prediction from slickwater/linear gel treatment, which is scaled down to ~2/3 of the initial prediction to account for the heterogeneity of the reservoir model due to a simplified reservoir model used for pre-stimulation condition. The blue curve is the production from conventional two-wing gel fracturing design. It is found that post-frac flow rate from slickwater stimulation is expected to be about three times the flow rate from the gel treatment in the stabilized regime (one year after stimulation). Although actual flow rates from both treatments depends on the applied drawdown, the corresponding flow rates after one year are expected to be ~55 × 10^4 m^3/d for slickwater treatment and ~ 17 × 10^4 m^3/d for gel treatment, respectively, with a constant drawdown of 20 MPa.

Optimizing Hydraulic Fracturing Treatment Integrating Geomechanical Analysis and Reservoir
Simulation for a Fractured Tight Gas Reservoir, Tarim Basin, China

197

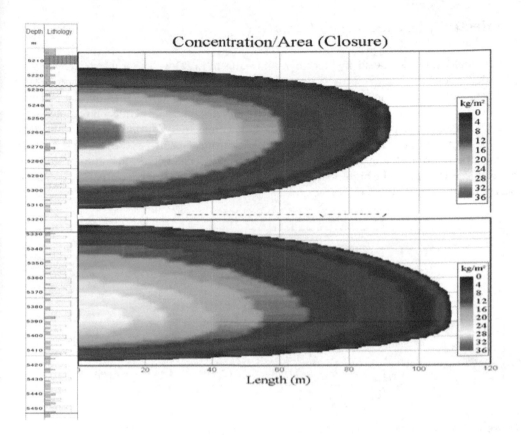

Figure 15. Fracture growth and proppant coverage (colour) for two stage of hydraulic fractures using conventional gel treatment.

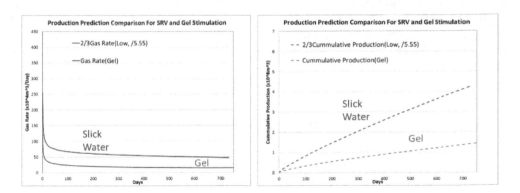

Figure 16. Production prediction comparison of two different hydraulic fracturing treatments. The red curve is the production prediction from slickwater/linear gel treatment; the blue curve is the production from conventional two-wing gel fracturing design.

7. Injectivity test and stage 1 treatment

It was decided to test the slickwater/liner gel treatment in D3 well after the study was completed. A pre-stimulation injectivity test was performed through perforations prior to Stage 1 and after the mini-frac test (Figure 17). Interestingly, the test showed the opposite behavior from what one would expect if the stimulation enhances reservoir permeability. Later-stage injectivity (during step-down) is lower than early stage injectivity (during step-up), rather than higher. Although there might be other reasons affect the test result, i.e., the un-stable injection during the whole test, it is believed the main reason was lack of access to natural fractures in the tested interval and the high closure pressure because the test was conducted in a cased and perforated hole and after a mini-frac.

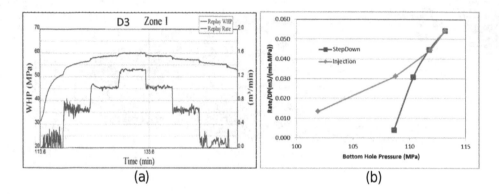

(a)   (b)

Figure 17. Pre-stimulation injectivity test pressure curve (a) and injectivity interpretation (b).

The Stage 1 treatment was conducted using slickwater and linear gel after the injectivity test. However, a screen out was experienced at the end of the execution and tubing leakage was discovered afterwards. Treatments in the other three zones had not occurred at the date of writing this paper. The stage 1 production test is still very promising, and it has been decided to continue slickwater/linear gel treatment in other three stages after the tubing problem is fixed.

8. Discussion and conclusion

In this paper we have outlined a new workflow for simulation of a complex fracture network created by stimulation using low-viscosity fluids in a fractured tight sandstone reservoir. The workflow is based on critically stressed fracture theory. This process of natural fracture stimulation is believed to be the underlying reason for the success in shale gas reservoir stimulation. The results suggested that there would be significantly higher production from this approach compared to conventional two-wing gel fracturing.

Optimizing Hydraulic Fracturing Treatment Integrating Geomechanical Analysis and Reservoir
Simulation for a Fractured Tight Gas Reservoir, Tarim Basin, China

199

There are, however, some uncertainties in the modeling of the natural fracture stimulation for this fractured tight gas reservoir.

1. The pressure-permeability relationship used in modeling the permeability enhancement by slickwater stimulation is taken from a shale gas field. It is unclear whether the data from the analogue field drilled through mudstones will be applicable to the modeled fractured tight sandstone reservoir. Post-stimulation production simulation, or a pre-stimulation injectivity test in nearby wells in open hole could help to better constrain this relationship, hence improve the accuracy of the prediction.

2. Due to the lack of knowledge of fracture distribution between wells, the fractures interpreted from all three offset wells were used to predict the stimulation behavior of natural fractures, and it was assumed that a similar fracture distribution would be found in all formations. In reality, the fracture distribution is likely to be different, depending among other things on the lithology and structural location. For example, it is already noticed that there are fewer fractures in the lower part of the reservoir than in the upper part in the D3 well. Intervals with dense fracture networks are more likely to benefit from slickwater treatment compared to formations with no or very sparse fractures. A 3D description of the fracture distribution is always preferred.

3. Micro-seismic imaging is not available in the study area. No wells are close enough to work as a monitoring well and surface monitoring is also impossible due to the great depth of the reservoir. The lack of microseismic data made it impossible to calibrate the prediction of the shape of SRV.

The main uncertainty in gel frac productivity estimation comes from the propped fracture conductivity estimation. This conductivity is based on proppant testing in the laboratory. The proppant inside fractures involves clogging, crashing and embedment over the production period. There is no analytical method available to model these long-term effects on propped fracture conductivity. An approximate conductivity damage factor has been used in this study to consider these effects.

Although there are still some shortcomings with the workflow, it can assist in the assessment of development concepts and the evaluation of stimulation enhancement options. The anisotropy in the slickwater treatment can be reasonably well-predicted and applied into the production simulation, which provides a more robust prediction than a simple isotropy model. The new workflow can be used in naturally fractured shale gas, tight gas/oil and CBM reservoirs.

Acknowledgements

The authors wish to thank PetroChina Tarim Oil Company for providing us with the data and for permission to publish this paper, and Baker Hughes internal support to carry out the work.

Author details

Feng Gui[1*], Khalil Rahman[1], Daniel Moos[2], George Vassilellis[3], Chao Li[3], Qing Liu[4], Fuxiang Zhang[5], Jianxin Peng[5], Xuefang Yuan[5] and Guoqing Zou[5]

*Address all correspondence to: Feng.gui@bakerhughes.com

1 Baker Hughes, Perth, Australia

2 Baker Hughes, Menlo Park, USA

3 Gaffney, Cline &Associates, Houston, USA

4 Baker Hughes, Beijing, China

5 PetroChina Tarim Oil Company, Korla, China

References

[1] Modeland, N, Buller, D, & Chong, K. K. Stimulation's influence on production in the Haynesville Shale: a playwide examination of fracture-treatment variables that show effect on production. In: proceedings of Canadian Unconventional Resources Conference, CSUG/SPE November (2011). Calgary, Alberta, Canada., 148940, 15-17.

[2] Maxwell, S. C, Pope, T, Cipolla, C, et al. Understanding hydraulic fracture variability through integrating microseismicity and seismic reservoir characterization. In: proceedings of SPE North American Unconventional Gas Conference and Exhibition, SPE June (2011). Woodlands, Texas, USA., 144207, 14-16.

[3] Sayers, C. and Le Calvez, J., (2010). Characterization of microseismic data in gas shales using the radius of Gyration tensor, SEG Expanded Abstract.

[4] Moos, D. Improving Shale Gas Production Using Geomechanics, Exploration & Production- Oil & Gas Review (2011). , 9(2), 84-88.

[5] Zoback, M. D, Kohli, A, Das, I, & Mcclure, M. The importance of slow slip on faults during hydraulic fracturing stimulation of shale gas reservoirs. In: proceedings of SPE Americas Unconventional Resources Conference, SPE June (2012). Pittsburgh, Pennsylvania, USA., 155476, 5-7.

[6] Mullen, M, & Enderlin, M. Is that frac job really breaking new rock or just pumping down a pre-existing plane of weakness?- the integration of geomechanics and hydraulic-fracture diagnostics. In: proceedings of 44th US Rock Mechanics Symposium and 5th US-Canada Rock Mechanics Symposium, ARMA 10-285, 27-30 June (2010). Salt Lake City, UT, USA.

Optimizing Hydraulic Fracturing Treatment Integrating Geomechanical Analysis and Reservoir
Simulation for a Fractured Tight Gas Reservoir, Tarim Basin, China

201

[7] Moos, D, Vassilellis, G, & Cade, R. Predicting shale reservoir response to stimulation in the Upper Devonian of West Virginia. In: proceedings of SPE Annual Technical Conference and Exhibition, SPE-145849, 30 October-2 November (2011). Denver, Colorado, USA.

[8] Vassilellis, G. D, Li, C, Moos, D, et al. Shale engineering application: the MAL-145 Project in West Virginia. In: proceedings of Canadian Unconventional Resources Conference, CSUG/SPE-November (2011). Calgary, Alberta, Canada., 146912, 15-17.

[9] Barton, C, Zoback, M. D, & Moos, D. Fluid Flow Along Potentially Active Faults in Crystalline Rock, Geology(1988). , 23(8), 683-686.

[10] Moos, D, & Barton, C. A. Modeling uncertainty in the permeability of stress-sensitive fractures. In: proceedings of 42nd US Rock Mechanics Symposium and 2nd U.S.-Canada Rock Mechanics Symposium, ARMA June- 2 July (2008). San Francisco, USA., 08-312.

[11] Hossain, M. M, Rahman, M. K, & Rahman, S. S. A Shear Dilation Stimulation Model for Production Enhancement From Naturally Fractured Reservoirs, SPE 78355, SPE Journal; June , 2002-183.

[12] Moos, D, & Zoback, M. D. Utilization of Observations of Well Bore Failure to Constrain the Orientation and Magnitude of Crustal Stresses: Application to Continental, Deep Sea Drilling Project and Ocean Drilling Program Boreholes, Journal of Geophysical Research (1990). , 95, 9-305.

[13] Heidbach, O, Tingay, M, Barth, A, Reinecker, J, Kurfe, D, & Müller, B. The World Stress Map database release (2008). doi:10.1594/GFZ.WSM.Rel2008.http://www.world-stress-map.org

Permissions

The contributors of this book come from diverse backgrounds, making this book a truly international effort. This book will bring forth new frontiers with its revolutionizing research information and detailed analysis of the nascent developments around the world.

We would like to thank the Organising Committee, for lending their expertise to make the book truly unique. They have played a crucial role in the development of this book. Without their invaluable contribution this book wouldn't have been possible. They have made vital efforts to compile up to date information on the varied aspects of this subject to make this book a valuable addition to the collection of many professionals and students.

This book was conceptualized with the vision of imparting up-to-date information and advanced data in this field. To ensure the same, a matchless editorial board was set up. Every individual on the board went through rigorous rounds of assessment to prove their worth. After which they invested a large part of their time researching and compiling the most relevant data for our readers. Conferences and sessions were held from time to time between the editorial board and the contributing authors to present the data in the most comprehensible form. The editorial team has worked tirelessly to provide valuable and valid information to help people across the globe.

Every chapter published in this book has been scrutinized by our experts. Their significance has been extensively debated. The topics covered herein carry significant findings which will fuel the growth of the discipline. They may even be implemented as practical applications or may be referred to as a beginning point for another development. Chapters in this book were first published by InTech; hereby published with permission under the Creative Commons Attribution License or equivalent.

The editorial board has been involved in producing this book since its inception. They have spent rigorous hours researching and exploring the diverse topics which have resulted in the successful publishing of this book. They have passed on their knowledge of decades through this book. To expedite this challenging task, the publisher supported the team at every step. A small team of assistant editors was also appointed to further simplify the editing procedure and attain best results for the readers.

Our editorial team has been hand-picked from every corner of the world. Their multi-ethnicity adds dynamic inputs to the discussions which result in innovative outcomes. These outcomes are then further discussed with the researchers and contributors who give their valuable feedback and opinion regarding the same. The feedback is then collaborated with the researches and they are edited in a comprehensive manner to aid the understanding of the subject.

Apart from the editorial board, the designing team has also invested a significant amount of their time in understanding the subject and creating the most relevant covers. They scrutinized every image to scout for the most suitable representation of the subject and create an appropriate cover for the book.

The publishing team has been involved in this book since its early stages. They were actively engaged in every process, be it collecting the data, connecting with the contributors or procuring relevant information. The team has been an ardent support to the editorial, designing and production team. Their endless efforts to recruit the best for this project, has resulted in the accomplishment of this book. They are a veteran in the field of academics and their pool of knowledge is as vast as their experience in printing. Their expertise and guidance has proved useful at every step. Their uncompromising quality standards have made this book an exceptional effort. Their encouragement from time to time has been an inspiration for everyone.

The publisher and the editorial board hope that this book will prove to be a valuable piece of knowledge for researchers, students, practitioners and scholars across the globe.

List of Contributors

Hyunil Jo and Robert Hurt
Baker Hughes, Tomball, TX, USA

B. Damjanac, C. Detournay, P.A. Cundall and Varun
Itasca Consulting Group, Inc., Minneapolis, Minnesota, USA

R. G. Jeffrey and Z. Chen
CSIRO Petroleum and Geothermal, Australia

K. W. Mills
SCT Operations Pty Ltd, Australia

S. Pegg
Narrabri Coal Operations Pty Ltd, Australia

Smarajit Sengupta, Dhubburi S. Subrahmanyam, Rabindra Kumar Sinha and Govinda Shyam
National Institute of Rock Mechanics, Bengaluru, India

Somayeh Goodarzi and Antonin Settari
University of Calgary, Calgary, Canada

Mark Zoback
Stanford University, USA

David W. Keith
Harvard University, USA

Bisheng Wu, Xi Zhang and Rob Jeffrey
CSIRO Earth Science and Resource Engineering, Melbourne, Australia

Andrew Bunger
CSIRO Earth Science and Resource Engineering, Melbourne, Australia
Department of Civil and Environmental Engineering, University of Pittsburgh, Pittsburgh, PA, USA

Luke Frash, Marte Gutierrez and Jesse Hampton
Colorado School of Mines, Golden, CO, USA

Sergey Turuntaev and Evgeny Zenchenko
Institute of Geosphere Dynamics of Russian Academy of Sciences (IDG RAS), Moscow, Russia

Olga Melchaeva
Moscow Institute of Physics and Technology, Moscow, Russia

Michael Molenda, Ferdinand Stöckhert, Sebastian Brenne and Michael Alber
Ruhr-University Bochum, Germany

Reza Keshavarzi and Reza Jahanbakhshi
Young Researchers and Elites Club, Science and Research Branch, Islamic Azad University, Tehran, Iran

Feng Gui and Khalil Rahman
Baker Hughes, Perth, Australia

Daniel Moos
Baker Hughes, Menlo Park, USA

George Vassilellis and Chao Li
Gaffney, Cline & Associates, Houston, USA

Qing Liu
Baker Hughes, Beijing, China

Fuxiang Zhang, Jianxin Peng, Xuefang Yuan and Guoqing Zou
PetroChina Tarim Oil Company, Korla, China

Printed in the USA
CPSIA information can be obtained
at www.ICGtesting.com
JSHW011406221024
72173JS00003B/434

9 781632 394248